Economic Development
OF LOW EARTH ORBIT

Edited by Patrick Besha and Alexander MacDonald

National Aeronautics and Space Administration

NASA Headquarters
300 E Street SW
Washington, DC 20546

NP-2016-03-2140-HQ

Table of Contents

PREFACE v

Patrick Besha and Alexander MacDonald, Editors

CHAPTER 1 1

Selecting Policy Tools to Expand NASA's Contribution to Technology Commercialization

Gregory Tassey

CHAPTER 2 23

Protein Crystallization for Drug Development: A Prospective Empirical Appraisal of Economic Effects of ISS Microgravity

Nicholas S. Vonortas

CHAPTER 3 43

Does Information About Previous Projects Promote R&D on the International Space Station?

Albert N. Link and Eric S. Maskin

CHAPTER 4 61

Venture Capital Activity in the Low-Earth Orbit Sector

Josh Lerner, Ann Leamon, and Andrew Speen

CHAPTER 5 113

Directing vs. Facilitating the Economic Development of Low Earth Orbit

Mariana Mazzucato and Douglas K. R. Robinson

ACRONYMS 131

Preface

Patrick Besha, Editor, Senior Policy Advisor, NASA
Alexander MacDonald, Editor, Senior Economic Advisor, NASA

IN THE NEXT DECADE, NASA will seek to expand humanity's presence in space beyond the International Space Station (ISS) in low Earth orbit to a new habitation platform around the Moon. By the late 2020s, astronauts will live and work far deeper in space than ever before. As part of our push outward into the solar system, NASA is working to help commercialize human spaceflight in low Earth orbit. After the government pioneers, develops, and demonstrates a space capability—from rockets to space-based communications to Earth observation satellites—the private sector realizes its market potential and continues innovating. As new companies establish a presence, the government often withdraws from the market or becomes one of many customers.

In 2016, we are once again at a critical stage in the development of space. The most successful long-term human habitation in space, orbiting the Earth continuously since 1998, is the ISS. Currently at the apex of its capabilities and the pinnacle of state-of-the-art space systems, it was developed through the investments and labors of more than a dozen nations and is regularly resupplied by cargo delivery services. Its occupants include six astronauts and numerous other organisms from Earth's ecosystems, from bacteria to plants to mice. Research is conducted on the spacecraft from hundreds of organizations worldwide, ranging from academic institutions to large industrial companies and from high-tech start-ups to high school science classes. However, its operational lifetime may be exceeded by the late 2020s, compelling its retirement to make way for new spacecraft and new missions.

As NASA begins moving astronauts out to the lunar vicinity, Mars, and beyond, the Agency will leave the further development of low Earth orbit to private sector companies. This has the potential to be a historic transition—from a government-run laboratory in orbit to an independent human spaceflight economy.

In order for a viable, sustainable economy based on human spaceflight to emerge in low Earth orbit (LEO), a number of elements must be present. First, the marketplace dynamics of supply and demand must exist. Second, the overwhelming reliance on government demand and public procurement must be transitioned to a market in which industry and other private sector demand is the primary market force, met by industry supply. The transition from government-led to private sector–led human spaceflight activity in LEO will constitute a great experiment in the development of American spaceflight capabilities, and the careful management of the dynamics of this transition will be of paramount importance.

NASA has taken a number of productive steps to support the fledgling commercial human spaceflight industry, including the creation of several programs aimed at supporting private sector firms' development of essential space infrastructure and transportation. Examples include the Commercial Orbital Transportation Services (COTS) program, which funded both SpaceX and Orbital ATK (formerly Orbital Sciences) to develop the capability to ferry cargo from Earth to the ISS. Building on its success, NASA then awarded commercial resupply services contracts to these providers and initiated the Commercial Crew Program, which is currently funding Boeing and SpaceX to develop spacecraft capable of transporting astronauts to the ISS. In 2015, NASA awarded a second round of resupply contracts to SpaceX and Orbital ATK, as well as a new provider, the Sierra Nevada Corporation. As contracted commercial suppliers, these companies, and the Commercial Crew Program companies, will also have the legal right to sell flights of their vehicles to other customers, opening up opportunities for broader LEO commercialization.

Similarly, NASA established the Center for the Advancement of Science in Space (CASIS) in 2011 to be the manager of the ISS National Laboratory. Given direction to fund commercial R&D, the Center has seeded dozens of projects that have flown in space. As the primary portal for companies interested in utilizing the ISS, CASIS is crucial to expanding private sector interest in LEO.

Recent developments in spaceflight suggest there is ample cause to be optimistic about the future. The next generation of habitation modules, such as those that can operate in low Earth orbit and also around the Moon, are currently under development. In 2016, Bigelow Aerospace is slated to dock its experimental prototype module to the ISS in a first-of-its-kind demonstration and a clear signal that the beginning of the ISS transition era is upon us.

Furthermore, the landings of reusable rockets by SpaceX and Blue Origin represent a groundbreaking milestone in the history of spaceflight. In addition to greatly advancing the state of rocketry, the new capability may have a significant democratization and commercialization effect, potentially enabling low-cost access to space for entrepreneurs, scientists, educators, and the general public.

As the overall strategy for the economic development of LEO emerges, NASA asked a small group of prominent economists to examine some of the most important questions facing the Agency as it enters into this historic transition. These papers provide independent perspectives that do not necessarily reflect NASA policy but which we find to be valuable in raising important issues and asking challenging questions.

In order to stimulate demand-side LEO commercialization activities effectively, the Agency will need a policy road map to make the best technology development decisions. **Gregory Tassey** offers a complex but logical path to success by outlining a system of policies based on the level and breadth of technology platforms desired. Such a plan could be used to implement elements of an active innovation policy and to integrate LEO activities more closely into the national innovation system.

Nicholas Vonortas examines a crucial piece of this puzzle: what intrinsic qualities of space enable and support economic activity? He finds that the unique microgravity environment of space is perhaps its greatest untapped value. One of the most likely beneficiaries of microgravity research may be the biotech industry. A promising line of research suggests that the microgravity environment enables protein crystals to be grown significantly better than in terrestrial laboratories. Such crystals play a fundamental role in pharmaceutical development. But how can we measure the additionality of microgravity? How might it improve pharmaceutical development? The paper provides a practical application of economic theory to a vexing measurement problem in the emerging LEO economy.

What are the costs, both in time and money, associated with commercial operations in space and how does knowing—or not knowing—that information affect investment decisions? To answer this crucial question, **Albert Link and co-author Eric Maskin**, a Nobel prize-winning economist, consider the current R&D environment on the ISS, with a goal to offer policy suggestions for improvement. They find that a lack of information about past projects, experimental success rates, and the flight process in general were major factors inhibiting both R&D and commercial growth. Without such information, researchers and companies were unable to accurately assess the risks involved. The solution? An easily searchable, highly transparent database could provide the necessary information to lower the barrier to entry for commercial operations in space.

How will innovative companies emerge? **Josh Lerner, Ann Leamon, and Andrew Speen** present a detailed examination of venture capital (VC) interest in the sector. While investors are perhaps understandably less cognizant of the opportunities LEO offers, VC may be an important source of funding for early-stage companies once the market matures. Furthermore, the emergence of significant near-term start-up successes that utilize human spaceflight capabilities in LEO could spur increased VC investment in the sector.

To close out the collection, **Mariana Mazzucato and Douglas Robinson** highlight some of the challenges associated with facilitating and directing development and suggest that NASA should seek to foster a robust innovation and industrial policy ecosystem to achieve mission-focused goals in LEO. Such goals would include NASA being at the forefront of strategic, high-risk investments in the near-term and channeling any resulting technology or knowledge to the private sector to spur economic growth.

This collection of papers identifies a number of important policy questions that will be of rising importance as NASA transitions human spaceflight in LEO to the private sector, as well as a number of economic analysis methods for addressing those questions. Life off of the Earth is a new field of social and economic organization that will have vast implications for our evolution and our future. Economic development in orbit is necessary for that future growth. It is our hope that this volume may serve to guide decisions and spark the intellectual curiosity of space policy makers, NASA program managers, economic researchers, and all others interested in the continued economic development of human spaceflight.

CHAPTER 1

Selecting Policy Tools to Expand NASA's Contribution to Technology Commercialization

Gregory Tassey

ORIGINALLY CHARTERED TO ADVANCE the U.S. science and technology enterprise, NASA's efforts are strengthened by its enduring relevance to the national economy. The Agency plays a role in developing core technology platforms and supporting technical infrastructures upon which many applied technologies can be built. While NASA may primarily serve as a foundational technology development agency, its investments and policies should consider the entire technology lifecycle, including commercial applications that provide desired economic benefits. In order for the LEO commercialization endeavor to be successful it will need to contribute to the Nation's economic growth. Doing so will require effective management of technology investment.

Why is investment in technology important? Consider that in spite of a modest economic recovery, including increases in employment, median real household income is 9 percent lower than in 2000. In addition, income inequality has reached the historic highs set in the late 1920s. Consequently, U.S. policy makers have become increasingly concerned about the lack of wage and income growth for the majority of Americans. Polls show that the American public ranks "jobs" as its number 1 concern.

The reason for low income growth is a lack of sustained investment in productivity-enhancing assets: technology, hardware and software that embody new technology, skilled labor to use the new technology, and technical infrastructure that enables

Disclaimer: The views and opinions of the authors do not necessarily state or reflect those of the U.S. Government or NASA.

the technology-based economic process.[1] Economic studies show that the technology-based economy not only enables the realization of social goals (national security, energy independence, environmental quality, space exploration, and health) but also enables higher overall productivity. The economic impact of productivity growth is the creation of higher paying jobs. BLS data show that in all but one of 71 technology-oriented occupations, the median income exceeds the median for all occupations. Moreover, in 57 of these occupations, the median income is 50 percent or more above the overall industry median.[2]

However, modern technologies are complex systems of hardware and software. This means that a number of technology trajectories must be initiated in parallel and managed efficiently to achieve high rates of innovation in the shorter times required by increasing global competition. The imperative to expand the high-tech economy has pushed industrialized nations to invest in steadily higher amounts of research and development (R&D). The world now spends $1.5 trillion per year on R&D. While this is a large investment by itself, it is only the tip of the global economic iceberg. Every dollar of R&D spawns many dollars of subsequent capital investment, manufacturing, and marketing activity.

More recently, governments have begun to experiment with new research and commercialization infrastructures to better manage the composition of R&D over a technology's development cycle and to attain greater efficiency with respect to both the rate and breadth of innovation.[3] The evolution of R&D portfolio management techniques, research consortia, "innovation clusters," incubators and accelerators, and national research and testing facilities have proliferated across the world's economy to improve the efficiency dimension.

The pursuit of R&D efficiency is resulting in significant "institutional" innovations. A major policy thrust in this regard has been the creation of new research entities such as the National Network for Manufacturing Innovation (NNMI), which involves cooperative investment by both the public and private sectors at regional Manufacturing Innovation Institutes (MII). The objective is to combine research

---

1 Gregory Tassey, "Why the US Needs a New Tech-Driven Growth Strategy," *Christian Science Monitor*, February 8, 2016 (*http://www.csmonitor.com/Technology/Breakthroughs-Voices/2016/0208/Why-the-US-needs-a-new-tech-driven-growth-strategy*). The full report is available from the Information Technology and Innovation Foundation Web site.

2 Daniel Hecker, "High-Technology Employment: A NAICS-based Update," *Monthly Labor Review* (July 2005): 57–72.

3 Gregory Tassey, "Beyond the Business Cycle: The Need for a Technology-Based Growth Strategy," *Science and Public Policy* 40:3 (2013): 1–23, and "Competing in Advanced Manufacturing: The Need for Improved Growth Models and Policies," *Journal of Economic Perspectives* 28:1 (Winter 2014): 27–48.

assets from public and private sources and thereby more efficiently advance new manufacturing technologies.[4]

As part of this major adjustment to economic growth strategy, the National Laboratory System is being pushed to increase its efficiency in not only delivering technologies that meet the respective parent agencies' missions but also to contribute to the increasingly important "dual use" mandate. Specifically, this mandate emphasizes (1) the spinoff/transfer of agency technologies to the private sector for further development and eventual commercialization, and (2) participation in the development of new commercial technologies through deployment of the agency's unique research capabilities in partnership with private companies.

In summary, the new institutional innovations are characterized by the integration of the multiple technology-related assets developed through complementary public-private investment strategies. Such integration increases the productivity of R&D, the diffusion of technical knowledge, and the efficiency of scale-up for initial production.

Section 1. Toward a NASA Technology Commercialization Strategy

One of the prominent categories of research assets in the United States is the National Laboratory System.[5] However, having been developed to meet a number of critical national objectives, these laboratories have pursued a set of technology trajectories determined by development and utilization criteria adapted for each agency's specific mission. In many cases, an agency is the consumer of its technology program's output. Examples are the Department of Defense (DOD) and the National Aeronautics and Space Administration (NASA). This "closed" R&D system leads to management methods that are optimized for Agency use of the resulting technology, but which are different from those required for technology commercialization objectives that respond to the national economic growth and competitiveness imperative.

Globally, governments have been restructuring their national labs to be more effective in leveraging technology assets in support of economic growth. In effect, U.S. R&D agencies are being directed to develop and implement dual use strategies for their laboratories and other institutional assets. As indicated above, the technology

---

4 Five MII have been formed in the areas of additive manufacturing, digital manufacturing and design, lightweight innovations, energy (NBG semiconductors), and advanced composites manufacturing. Three other MII are in the startup phase; they will focus on flexible hybrid electronics, integrated photonics, and clean energy manufacturing. See *http://www.manufacturing.gov/nnmi.html*.

5 There are 22 laboratories in the "National Laboratory System," 17 of which are operated by the Department of Energy. One of the four non-DOE labs is NASA's Center for the Advancement of Science in Space, operated within the International Space Station. See *http://en.wikipedia.org/wiki/United_States_national_laboratories*.

commercialization objective is forcing adoption of a management strategy driven by the three major R&D metrics identified above: amount, composition, and efficiency.

In this context, NASA management has initiated a process to develop a strategy for expanding and effectively utilizing its R&D assets to help meet the national goal of increased LEO commercialization. While NASA has considerable and varied R&D assets, the potential role of the International Space Station (ISS) has attracted attention as a potential comparative advantage for NASA based on the expectation that the uniqueness of a microgravity environment has abundant potential for the development of superior technologies, especially where molecular structure and its manipulation is a critical factor in a product technology's development and subsequent manufacture. Early efforts have focused on conducting scientific experiments in areas such as protein crystallization, where the microgravity environment has enabled the production of larger and purer crystals. This, in turn, increases the efficiency of subsequent research into protein drug development.[6]

However, a singular focus on the use of the ISS for early-phase research will likely underutilize NASA's much greater resources, which can potentially contribute to multiple phases of the R&D process beyond basic research and do so sooner in time. The follow-on phases of technology research—proof-of-concept research (technology platform development), subsequent applied research, and then the final development phase that leads directly to proprietary technology commercialization and hence economic growth—require a coordinated investment and an advanced R&D management infrastructure.

Although some future technologies may be amenable to full technology development and even manufacturing in low Earth orbit (LEO), the current experience is largely limited to scientific studies. Because of the multi-phased development path from scientific discovery to eventual market-ready technology, follow-on R&D and eventual manufacturing will likely be terrestrial based for some time. NASA therefore should consider a structured full-cycle R&D process strategy to significantly increase the probability of eventual commercialization and hence effective utilization of its considerable R&D resources.

For NASA to explore significant dual use of its major research facilities, it needs a comprehensive innovation framework that moves beyond traditional mission-oriented R&D management practices to a set of market-oriented policy tools that recognize private-sector barriers to investing in the several phases of the R&D process.

In response to the above economic trends and the mandate to more fully utilize U.S. government R&D assets, a policy model is presented for NASA management to use in supporting the development and commercialization of advanced technologies

---

6 An example is NASA's support of protein crystallization research using their National Laboratory (International Space Station (ISS). See *http://www.spaceflorida.gov/news/2012/11/01/casis-announces-first-grants-for-protein-crystallization* and *http://www.iss-casis.org/NewsEvents/PressReleases/tabid/111/ArticleID/49/ArtMID/586/CASIS-Announces-First-Grants-for-Protein-Crystallization.aspx#sthash.EubPuDfd.dpuf*.

that benefit from the unique characteristics of the International Space Station (ISS) and other R&D assets currently within NASA or from assets that could be acquired by NASA as part of a comprehensive technology commercialization strategy.

By focusing on the major elements that make up modern technologies, the policy model can enable NASA management to identify and characterize underinvestment across the R&D process and thereby drive the selection of appropriate policy instruments for supporting each phase of the target technology's life cycle. This broader and structured approach should increase expectations for and the probability of earlier success of NASA's initiatives to create a competitive institutional ecosystem for advanced R&D and subsequent manufacturing innovation.

In summary, having a national laboratory with unique research facilities as part of the ISS plus a broad technology development capability, NASA should consider pursuing opportunities to (1) expand the use of the ISS's unique research capabilities, (2) participate in early- and mid-R&D process institutions (MII), and (3) support commercialization efforts by companies participating in an MII and other "innovation cluster" related infrastructures.

Section 2. The Evolving Role of National Laboratories

A major feature of the evolving technology-based growth model pursued by economies around the world is the integration of national laboratories into the technology life cycle. For many decades, national laboratories were largely independent research entities whose research results were diffused/transferred at arm's length to other research institutions focusing on later phases of the R&D process and eventual commercialization.

However, industrialized nations have begun to address the need for greater R&D efficiency in response to a relentless expansion of the technology-driven global economy by broadening the roles of and more closely integrating their national laboratories into their economies' innovation infrastructures.

Once thought of solely as a source of scientific advances, the effectiveness of a national laboratory is now being assessed using a broad range of metrics, including both science and technological advances based on utilization of unique research facilities and increasing rates of transfer of technology to the private sector for commercialization.

This emphasis on commercialization requires a comprehensive set of tools. Examples are:

- collaborative research
- broader support of innovation infrastructure (cluster models)
- user facilities at national laboratories for academic and private-sector researchers doing both product and process technology development

- licensing of internally developed technologies
- technical assistance to companies for follow-on technology development

In response, the U.S. National Laboratory System is changing. In October 2011, President Obama issued a memorandum directing agencies with federal laboratories to accelerate technology transfer and commercialization of research, and to take steps to increase partnerships between businesses and laboratories.[7]

The Department of Defense (DOD) and the Department of Energy (DOE) have taken steps to respond to this shift in national strategy. However, distinctions between the two agencies' efforts are important for NASA strategic planning relating to the technology commercialization objective. While DOD's rationale for a full technology life cycle strategy is straightforward—it is the user of the ultimate technologies developed—DOE's mission relies on a market-based based process in which industry must make applied R&D and subsequent commercialization decisions to achieve society's objectives of energy efficiency and clean energy. Thus, DOE was already more oriented toward understanding private sector investment incentives and incorporating them into its R&D management structure than other R&D agencies and was therefore both more able and motivated to expand its set of policy tools.

DOE's core role is primarily focused on early- and mid-phase technology development to which its laboratories make major contributions.[8] However, the agency does to varying degrees subsidize later-phase R&D and in some areas, such as solar energy and electric vehicles, it has provided subsidies for production scale-up. Congress added tax incentives to spur commercialization of solar energy. This expanding array of policy tools being implemented by DOE is the direct result of the need to combine complementary assets by government and industry to eventually achieve technology commercialization. NASA will have to adopt a similar public-private technology development model.

DOE describes its laboratories' roles in terms of a set of capabilities with broad applications:

- execute long-term government scientific and technological missions, often with complex security, safety, project management, or other operational challenges;

---

[7] "Presidential Memorandum—Accelerating Technology Transfer and Commercialization of Federal Research in Support of High-Growth Businesses" (*https://www.whitehouse.gov/the-press-office/2011/10/28/presidential-memorandum-accelerating-technology-transfer-and-commercialization* and *http://thehill.com/blogs/congress-blog/technology/216835-national-labs-play-unique-role-in-working-for-america*).

[8] The DOE laboratories are located across 14 states and employ 30,000 scientists. Once focused almost entirely on scientific research, DOE now views these labs as foundries of future technologies, occupying a research niche that universities and the profit-driven private sector cannot match. Examples of these laboratories' research topics include novel materials for lighter transportation vehicles, genomic tools that improve drought resistant crops, the Ebola virus, and cancer research. Paul Alivisatos, Dan Arvizu, Charlie McMillan and Terry Michalske, "National Labs Play a Unique Role in Working for America," *The Hill*, September 9, 2014.

- develop unique, often multidisciplinary, scientific capabilities beyond the scope of academic and industrial institutions, to benefit the nation's researchers and national strategic priorities; and

- develop and sustain critical scientific and technical capabilities to which the government requires assured access.

To help implement its technology commercialization strategy, DOE issued an *Agreements for Commercializing Technology (ACT)*—a set of guidelines to help companies bring new clean energy technologies and other innovations to the market faster through technology transfer and commercialization of research in cooperation with its laboratories.[9] To specifically help small businesses and entrepreneurs with limited resources and licensing experience, the agency issued a *Licensing Guide and Sample License* to increase the transparency of the licensing process and to explain the laws and policies governing the licensing of federally funded research, thereby helping reduce both time and cost to acquire intellectual property (IP) from DOE's Laboratories.[10]

Section 3. Issues for NASA

The broad scope of an R&D agency's role in developing technology for its core mission has led to detailed taxonomies for managing the R&D process. Originated by NASA and strongly embraced by DOD, this taxonomy consists of nine Technology Readiness Levels (TRLs).[11] Although the TRL's are not grouped by phase of R&D, their descriptions allow them to be grouped under the major phases of the R&D process as follows: TRL 1 is "scientific" research, TRLs 2 and 3 are "technology proof-of-concept" research (conducted largely in corporate central research laboratories, national laboratories and other research institutes), TRLs 4, 5, and 6 are the equivalent of "applied" research, and TRLs 7 and 8 are "development." TRL 9 is post-deployment or actual utilization experience (the equivalent of post-commercialization support), which drives ongoing improvements over a technology's life cycle.

However, the TRL structure only provides metrics for levels of technology development. This taxonomy may be sufficient when the agency is the user of the technology, which is the case for the DOD and NASA core missions. However, when the ultimate goal is innovation for commercial markets, this structure by itself is

---

9 Department of Energy, "Energy Department Announces New Initiative to Remove Barriers for Industry to Work with National Labs, Commercialize Technology," December 2011 (*http://energy.gov/articles/energy-department-announces-new-initiative-remove-barriers-industry-work-national-labs*).

10 See *http://technologytransfer.energy.gov/LicensingGuideFINAL.pdf*.

11 Various descriptions of the nine TRSs exist. See *http://en.wikipedia.org/wiki/Technology_readiness_level* for a description of the nine levels. NASA's characterization of the TRLs is summarized at *http://www.esto.nasa.gov/files/TRL_definitions.pdf* and *http://www.nasa.gov/content/technology-readiness-level*.

inadequate because it does not include or is not mapped on to industry investment assessments/metrics at each TRL. Industry investment in R&D is driven by technical and market risk vs. expected reward and their investment behavior therefore reflects both factors, which therefore must be taken into account for government policy tool selection.

The fact that DOE must rely on industry to progressively take over the R&D investment as the technology matures creates the need to map industry investment (or, more accurately, underinvestment) at each phase of the R&D process and to use such assessments to drive selection of policy mechanisms. As explicitly indicated by the following policy model, such strategies are complicated because simultaneous investment is required by both industry and government, especially during the early and middle stages of the R&D process. Joint conduct of the research, including within national laboratories, is a challenging management strategy.

To better focus the previous discussion with respect to the needed policy framework, the issues for NASA are to determine (1) how far forward in the R&D process to provide support for technology development and (2) what policy tools to use in doing so. These decisions will be driven by assessments of the desirability of a "dual-use" role for NASA assets, such as but not limited to the ISS, in order to more fully utilize NASA's unique capabilities. More specifically, a policy framework is mandatory to manage the R&D process in order to (1) select optimal R&D portfolios for the ultimate objective of technology commercialization, and (2) facilitate follow-on R&D and commercialization activities by industry. The bottom-line metric will be the ultimate return on investment and subsequent contribution to meeting national goals for increasing the economic benefits from federal investment in R&D.[12]

Early experiments on the ISS indicate that certain emerging technologies with substantial economic potential can be more efficiently advanced in a microgravity environment, specifically at the early technology development phase of the R&D process. Further experience will illuminate the scope and magnitude of the potential for the ISS to execute the role of a National Laboratory in facilitating downstream technology development and subsequent commercialization. Only additional properly constructed projects and accurate assessment of their results can determine if latter-phase R&D (development) and possibly manufacturing in cooperation with industry will be attractive in economic terms across a number of technologies initially supported by NASA.

---

12 To this end, the Department of Energy's Office of Energy Efficiency and Renewable Energy (EERE) has made a considerable effort to measure the downstream impacts of its supported research not just on subsequent R&D investment by industry but also on evolving industry structure, including small firm formation and supply-chain integration that constitute lasting economic impacts from the original R&D investments. See, for example, Gretchen Jordan et al, *A Framework for Evaluating R&D Impacts and Supply Chain Dynamics Early in the Product Life Cycle*. DOE, Office of Energy Efficiency and Renewable Energy, June 2014.

Section 4. The Technology Element Model for Driving Government Policy Management

A consensus conceptual model for selecting among the set of available policy instruments is needed to assure stakeholders (Congress and the White House) that the proposed support for private sector investment at each phase in the development and commercialization of new technologies is efficient in design and therefore likely to lead to positive results. Such a model also must provide a framework for designing and implementing an evaluation of the resulting economic impacts to respond to increasing demands for accountability and to enable future improvements in the policy tools.

Corporate strategic behavior clearly shows that technology investment is not homogenous and that technologies are not black boxes. Rather, the transition from basic science to commercial product requires development of three major elements: technology platforms, infratechnologies, and ultimately proprietary technologies. The two additional elements, "technology platforms" and "infratechnologies," exhibit degrees of "public good" content.

These elements are depicted in **Figure 1.1**. The red-shaded boxes represent "private" goods—meaning that private investment incentives are sufficient to create and

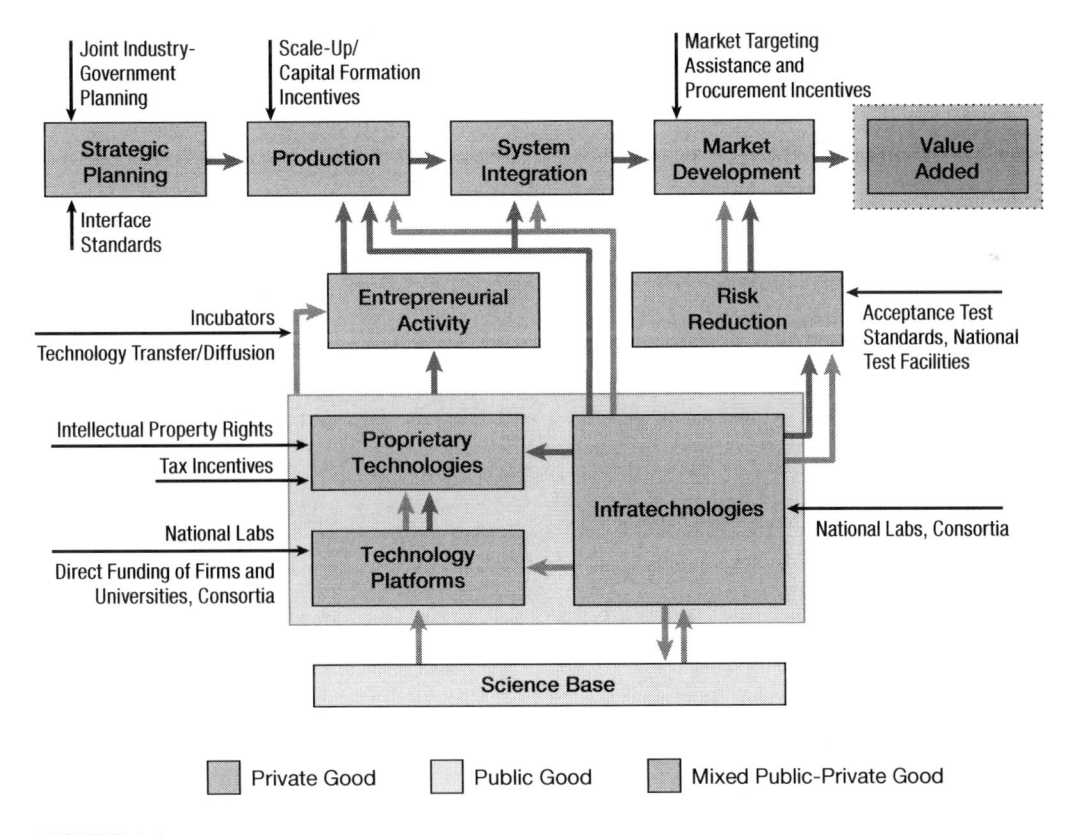

FIGURE 1.1: Managing the Entire Technology Life Cycle: Policy Roles in Response to Market Failures

use optimal amounts of these technology elements or activities. The blue-shaded boxes conversely represent "public" goods, whose character is such that industry substantially underinvests in them, leaving their funding largely, or at least partially, to government. Traditional economic growth models recognize basic science as a pure public good and Congress has a long history of funding scientific research. The policy problem arises in the boxes that are partially red and partially blue. These elements or activities are called "quasi-public" goods. As the label implies, their availability requires joint investment by public and private sources.

It is this last phenomenon that has led to efforts by technology-based economies to improve their growth models and thereby the efficiency of their domestic R&D through the creation of "innovation clusters" and broader and more effective use of national laboratories. The quasi-public technology elements or activities are the primary target for these policy initiatives. In the United States, the U.S. R&D agencies with the most advanced technological capabilities—DOD, DOE, and NASA, and NIST—are the logical agents of new policy initiatives.[13]

Within the R&D process, one of the most important policy targets is the "proof-of-concept" phase of research and development. The result of such research is the creation of broad technology-platforms, whose existence both confirms the potential for multiple market applications and provides a set of technical conceptualizations that drive the applied research and development leading to commercialization (i.e., innovations).

The existence of economies of scope, together with the high degree of technical and market uncertainty, means that individual firms will likely not capture many or most of the potential markets resulting from this type of research. However, large economies of scope (many potential market applications of the platform technology) are exactly what government support of technology commercialization wants to achieve because their existence means large potential economic impact. Such a result can only be achieved by enabling access to new technology platforms by many companies.

As an example, consider NASA's funding of a number of experiments using the ISS in the area of protein crystallization. This research is demonstrating that much larger and purer protein crystals can be grown in a microgravity environment. As large, high-quality crystals are essential to protein therapeutics design, the potential economic impact through eventual commercialization is significant. Moreover,

---

13 The attributes of such a model also include a time dimension; that is, management mechanisms are required to enable selection among suitable policy instruments as the technology evolves. The dynamic character of government support of the process of innovation makes this area of economic growth policy particularly difficult to manage, as private sector investment incentives continuously change over the R&D process and thus require continual shifts in the policy tool mix.

significant economies of scope are possible, given the large number of diseases potentially treatable by protein therapeutics.[14]

The implications for a NASA technology commercialization strategy are several. NASA could be satisfied with the singular contribution of increasing the productivity of university and biopharmaceutical company research with, say, the provision of particular biomolecules such as protein crystals. How and to what extent this advance in research capability/infrastructure affects future protein drug development would be left to the biopharmaceutical industry.[15]

Alternatively, NASA could support several early phases of the R&D process, beginning with proof-of-concept technology research for a new multidiscipline-based technology platform where NASA either applies other assets available within its R&D infrastructure or acquires them. This second option would require supporting a broad portfolio of research targeting the new technology platform that would enable a broad range of applications based in new protein therapeutic concepts.[16] The long-run goal would be substantial economic impact through achievement of economies of scope in market applications (many new drugs). A management strategy embodying tools for determining the appropriate R&D portfolio and then managing and coordinating the needed research at each phase of the R&D process would constitute the strategic scope of NASA support. Such a strategy is more complex and requires considerably more resources, but it is necessary if significant economic impact is to be realized. In the pharmaceutical industry, a technology platform involves a complete conceptual (proof-of-concept) model of the proposed new drug mechanism: specific biological targets, bioavailability, toxicity, and the like (see Table 1.1).

In contrast, smaller and ad hoc support of technology development is likely to yield little economic impact. In fact, even with considerable resources, the wrong policy model will likely yield poor results.

For several decades, the National Institutes of Health (NIH) spent billions of dollars on life-science research and then waited for private venture capital to fund the development of new biopharmaceuticals beginning with proof-of-concept/technology platform research. But although the basic science may be established, private risk

---

14 The protein therapeutics market includes peptide hormones, therapeutic enzymes, cytokines, monoclonal antibodies, blood factors, vaccines, and peptide antibodies. The global protein therapeutics market reached $138.3 billion in 2012 and is expected to increase to approximately $180 billion in 2018. Many companies participate in this market drawing upon existing technology platforms. Source: Boston Consulting Group (*http://www.giiresearch.com/report/bc63235-protein-drug.html*).

15 Protein drug development is used as an example because NASA is already supporting research in protein crystallization. Clearly a "lead agency" issue exists. However, if NASA has unique R&D assets appropriate for follow-on technology development or can rationalize acquiring them, supporting subsequent phases of the R&D process is justified.

16 For example, DOD's Defense Advanced Research Projects Agency (DARPA) has a project area entitled "Biological Technologies Office" (*http://www.darpa.mil/our_work/BTO*), which is targeting a set of technology trajectories aimed at integrating biology with other scientific disciplines to achieve radically new technology platforms.

TABLE 1.1: Application of the Technology-Element Model: Biotechnology

Science Base	Infratechnologies	Technology Platforms (Proof of Concept)		Commercial Products
		Product	**Process**	
Genomics	Bioinformatics	Antiangiogenesis	Cell encapsulation	Coagulation inhibitors
Immunology	Bioimaging	Antisense	Cell culture	DNA probes
Microbiology/virology	Biomarkers	Apoptosis	Microarrays	Inflammation inhibitors
Molecular and cellular biology	Combinatorial chemistry	Bioelectronics	Gene transfer	Hormone restorations
Nanoscience	DNA sequencing and profiling	Biomaterials	Gene testing	Nanodevices
Neuroscience	Electrophoresis	Biosensors	Immunoassays	Neuroactive steroids
Phramacology	Fluorescence	Functional genomics	Implantable delivery systems	Neurotransmitter inhibitors
Physiology	Gene expression analysis	Monoclonal antibodies	Nucleic acid amplification	Protease inhibitors
Proteomics	Magnetic resonance spectrometry	Pharmacogenomics	Recombinant DNA/genetic engineering	Vaccines
	Mass spectrometry	Stem cell	Separation technologies	
	Nucleic acid diagnostics	Tissue engineering	Transgenic animals gene delivery systems	
	Protein structure modeling and analysis techniques		Gene therapy	
			Gene expression systems	

Public Technology Goods → Mixed Technology Goods → Private Technology Goods

Source: Gregory Tassey, *The Technology Imperative*, Edwared Elgard, 2007.

capital does not like to assume the burden of developing new technology platforms due to the long time to market and the remaining high technical and market risk. In fact, over the last decade, venture capital has shifted toward later-stage product development (in this example, later-stage clinical trials). As a result, biopharmaceutical firms have attempted to develop new drugs directly from the underlying science, basically by using clinical trials (specifically, Phase II) to prove a new drug design. Phase II tests a specific drug design at considerable time at cost, but because this approach largely skips a true proof-of-concept phase, the probability of selecting a successful drug design is lowered. A study by the National Institute of Standards and Technology (NIST) found that this traditional drug development approach has led to low probabilities for drug candidates advancing through additional clinical testing and eventual approval by the Food and Drug Administration (FDA).[17]

Infratechnologies are the second of the quasi-public technology elements and suffers from similar public-private investment coordination, funding and conduct failures. The NIST study also examined the negative impact of inadequate infratechnologies in the biopharmaceutical industry. Based on industry surveys, it was estimated that more adequate and timely provisions of infratechnologies could lower the cost of taking a new drug candidate from conception to FDA approval by 25–48 percent and reduce the time to approval by 20 percent.

Infratechnologies are a diverse set of technical tools that are necessary to conduct all phases of research and development, to control production processes, and to execute marketplace transactions for complex technology-based goods. They include research tools such as measurement and test methods, scientific and engineering data, quality control techniques, and the functional as well as physical basis for the interfaces between components of modern technology systems. These tools are called "infratechnologies" because they provide a complex but essential technical infrastructure, which is as critical to achieving adequate private investment, and hence an adequate growth rate for the modern technology-based economy, as traditional economic infrastructure was for the Industrial Revolution.[18]

Proof-of-concept research, infratechnologies, and applied research and development exhibit distinctly different incentives and degrees of public good content. Each of these technology elements therefore requires a unique set of policy responses. Basic science is close to a pure public good, which is why it makes sense that the lion's share of basic research is funded by the government. Proof-of-concept (technology-platform) research and infratechnology research are typically co-funded by industry and government—hence the overall rationale for the evolving partnership mechanisms increasingly observed in the global manufacturing economy. The third element, proprietary technology, is closest to a pure private good, but even in this case, relatively high risk leads to underinvestment, which explains the existence of a "research and development" tax credit.

U.S. policy makers are finally responding by adopting R&D strategies that have already appeared in competing economies in Europe and Asia. National laboratories are increasingly making their unique research facilities available to companies to

---

17 Gallaher, M., J. Petrusa, A. O'Conner, and S. Houghton (2007). *Economic Assessment of the Technical Infrastructure Needs of the Biopharmaceutical Industry*. Gaithersburg, MD: National Institute of Standards and Technology (*http://www.nist.gov/director/planning/upload/report07-2.pdf*).

18 Infratechnologies are often embodied in standards that are ubiquitous in high-tech industries. The semiconductor industry has over 1,000 standards without which that industry could not function. Without the availability of this technical infrastructure (most of which is codified as industry standards), transaction costs would be higher not just at the research and development stage, but also during production and even marketing. Gallaher et al. (2007b) estimated that this industry spent $12 billion on measurement infratechnologies in the period 1996–2006, which generated gross benefits of $52 billion in 2006 dollars.

conduct proprietary research. Cooperative research centers are being established to combine public and private R&D assets to more efficiently achieve major advances in emerging technologies. Such public-private cooperative research efforts have finally been officially embraced by recent Congressional passage of legislation authorizing the National Network for Manufacturing Innovation (NNMI). An Advanced Manufacturing National Program Office (AMNPO) at NIST coordinates the growing number of Manufacturing Innovation Institutes (MII).

One of the major foci of the MII is nanotechnology. The same rationale for the technology element model applies here. In fact, investment in nanotechnology research is more coordinated and comprehensive in terms of investment in private and public elements than any previous national technology development effort (see **Table 1.2**).

TABLE 1.2: Application of the Technology-Element Model: Nanotechnology Platforms

Science Base	Infratechnologies	Technology Platforms (Proof of Concept) Product	Technology Platforms (Proof of Concept) Process	Commercial Products
Carbon-based nanomaterials	Biological detection and analysis tools	Carbon nanotubes	Epitaxy	Hardened nanomaterials for machining/drilling
Cellulosic nanomaterials	In silico modeling and simulation tools	Dendrimers	Nanoimprint lithography	Flame-retardant nanocoatings
Magnetic nanostructures	In-line measurement techniques to enable closed-loop process control	Hybrid nanoelectronic devices	Nanoparticle manufacture	Sporting goods
Molecular nanoelectronic materials		Ultra-low-power devices	Rapid curing techniques	Solar cells
Quantum dots		Self-powered nanowire devices	Self-assembling and self-organizing processes	Sunscreens/ cosmetics
Optical metamaterials	Sub-nanometer microscopy	Nanoparticle fluorescent labels for cell cultures and diagnostics	Scalable deposition method for polymer-fullerene photovoltaics	Targeted delivery of anticancer therapies
Solid-state quantum-effect nanostructures	High-resolution nanoparticle detection	Metal nanoparticles and conductive polymers for soldering/ bonding	Injet processes for printable elctronics	Biodegradable and lipid-based drug delivery systems
Functionalized fluorescent nanocrystals	Thermally stable nanocatalysts for high-temperature reactions		Purification of fluids with nanomaterials	Self-repairing and long-life wood composites
Quantum-confined structures		Nanoparticle sensors	Roll-to-roll processing	Antimicrobial coatings for medical devices
				Nanoscale motion microscopes

Public Technology Goods → Mixed Technology Goods → Private Technology Goods

Source: Gregory Tassey, *The Technology Imperative*, Edwared Elgard, 2007.

However, the NNMI program has recently been authorized by Congress but has not been funded directly. Instead, mission R&D agencies, such as DOD and DOE, have funded the MII that have already been established. Nevertheless, this approach could work if the portfolios of the MII are structured to ensure that the target technology platforms and supporting technical infrastructure are broad and deep enough to support follow-on applied R&D aimed at commercial markets. This situation is a major strategic issue for NASA management.

Going forward with a successful strategy requires a more accurate technology-element growth model that recognizes that growing technological complexity and the need to shorten R&D process times. Specific mechanisms are required to operationalize the characterization and assessment of the magnitude of the type of underinvestment occurring at each phase of R&D and subsequent commercialization. For example, at a specific phase of the R&D process, how can the policy process select between direct funding and the use of tax incentives? For underinvestment phenomena where direct funding is deemed the appropriate mechanism, how would NASA choose among options such as

- research within NASA laboratories;
- direct funding to individual companies;
- direct funding to research consortia; and
- promotion of full-scale innovation clusters.

Section 5. Summary of a Proposed NASA Strategy to Support Technology Commercialization

Because the majority of federal agency budgets support portfolios of R&D projects optimized for achieving their respective missions, programs aimed at economic growth as the final impact are served less well by traditional funding criteria. Thus, for an objective such as manufacturing in space for commercial purposes, an appropriate portfolio management framework must be developed using ongoing technology and economic assessments that reflect the commercialization objectives and the dynamic evolution of advanced manufacturing technologies. Such assessments include (1) regular projections of technical and market requirements, (2) analytical techniques that can determine the nature and magnitude of underinvestment at each phase of the technology life cycle, (3) the causes of underinvestment, (4) the selection of appropriate policy tools for each type of underinvestment, and, finally (5) an evaluation framework to collect impact data over the entire R&D process and subsequent commercialization.

The ultimate economic impact metric for technology platform development is the number and variety of products that can be developed from the underlying technology concept. The greater the "economies of scope," the greater the aggregate economic impact. An example involving NASA is "machine learning." Being developed to analyze enormous amounts of data now being generated from observations of the Milky

Way, the generic technology (technology platform) can be applied across a number of terrestrial applications, such as medical diagnosis and software engineering.[19]

Finally, it should be noted that this discussion is focused on the development of a single technology. Modern manufacturing technologies are systems of hardware and software. Developing each one of a technology system's components requires a multidisciplinary research effort over all phases of the R&D process. This is enough of a challenge, but then the technology system must be created through an integration process of all components, which requires functional standardized interfaces that are based on technologically sophisticated "infratechnologies."

For the potential productivity of such a "technology system" to be realized, all components must advance at some minimum rate for system productivity objectives to be realized. For example, automobiles used to be a modestly complex set of hardware components: engine, drive train, suspension, and the like. However, by 2011, the average automobile contained 17 subsystems for which electronics is a central element. These subsystems are controlled and connected to each other by nearly 100 microprocessors and five miles of wiring. Today, the number of processors has likely doubled.[20] Effective integration of the many hardware and software components into an efficient system is challenging and requires considerable system integration expertise, as well as an overall management structure that produce the desired system technology efficiently.

In this context, the following "policy elements" should be considered in formulating a technology commercialization:

1. Adopt the multi-element technology model to accurately define and manage the multiple government roles required to promote technology commercialization.

2. Emphasize "precompetitive" funding primarily through innovation clusters, which reduces IP issue (with respect to companies), increases research efficiency, and broadens the participating research base thereby accelerating technical knowledge diffusion/transfer.

3. Policy mechanisms for each phase of technology commercialization promotion pursued must be regularly assessed and their impacts monitored with respect to their appropriateness for each technology element and phase of R&D plus early commercialization efforts (especially scale-up to efficient production volumes); i.e., a real-time evaluation mechanism should be implemented.

---

19 NASA Jet Propulsion Laboratory, "Machines Teach Astronomers about Stars," January 8, 2015 (*http://www.jpl.nasa.gov/news/news.php?feature=4433* and *http://en.wikipedia.org/wiki/Machine_learning*).

20 International Center for Automotive Research. See Thomas R. Kurfess, "The Growing Role of Electronics in Automobiles: A Timeline of Electronics in Cars," 2011 (*http://www.chicagofed.org/digital_assets/others/events/2011/automotive_outlook_symposium/kurfess_060211.pdf*).

CHAPTER 1 ■ Selecting Policy Tools to Expand NASA's Contribution to Technology Commercialization

4. Label funded projects using the ISS and other laboratory research facilities as experimental/demonstration projects to more easily rationalize funding a large and broad portfolio, given the substantial risk associated with a new program.

5. Develop evaluation metrics for each phase of R&D supported projects to enable NASA management to track and assess progress and to make mid-course corrections as needed.

The various policy objectives, targets and mechanisms, and expected economic impacts associated with a comprehensive technology commercialization strategy are listed in Table 1.3 (p. 18). The critical overall message from the table is the need to first state the policy objective as a response to a specific market failure and then match the right policy tool with the nature of the underinvestment associated with a particular type of technology investment activity. These "policy tools" will consist of financial or in-kind support, where in-kind support includes the provision of government laboratories' unique research and testing facilities. As made clear in Table 1.3, the variety of economic activities and hence outcomes result in a number of impact metrics that can characterize the impact on economic growth.

The overall economic rationale for Table 1.3 is the fact that a set of substantial risks exists for technology investment that must be taken into account by government R&D funding programs. While economic growth impacts are larger from investments in technology than from any other asset, so-called "market failures" or underinvestment phenomena exist. Thus, investment incentives for technology are on average reduced by the fact that much of the required investment (R&D) must be undertaken years before initial commercialization with additional time required before major markets are established.[21]

Thus, even if the technology element model in Figure 1.1 (p. 9) is accepted by stakeholders, particularly Congress, and funding is initially provided, the long time required to develop a technology to the point of commercialization creates budget justification problems for the R&D agency. Congress funds agency R&D programs in one-year increments and expects evidence of progress in order to continue such funding. Obtaining funding for multiple years of research support for technologies projected to have economic growth impacts even farther in the future can thus be difficult.

---

21 When a technology is early in its development, both technical and market risks are at their highest levels. The corporate expected rate of return (ROR) calculation is lowered when the nominal expected ROR is reduced to account for these risks. The nominal expected ROR can be quite large for an entire industry and eventually even more so for the entire economy. However, the industry developing the new technology and especially individual companies in that industry can only expect to capture a fraction of the ultimate total economic benefits due a number of factors. This is a fundamental rationale for government R&D support. Moreover, the risk-adjusted expected ROR is further reduced for the time interval between R&D investment and commercialization, which can be years. Corporations apply a significant discount rate to the expected profits to account for the fact that the profits will be realized in the future. The discounting further discourages private-sector R&D investment.

TABLE 1.3: National Laboratories: Policy Needs, Mechanisms, and Implementation Targets for Technology Commercialization

POLICY NEED	POLICY OBJECTIVES/ TARGETS	IMPLEMENTATION MECHANISMS	ECONOMIC IMPACTS
R&D portfolios with broad commercial market potential	▪ Reorient research funding ▪ Support manufacturing and product technologies benefiting from unique federal research assets	▪ Add market potential assessments to research portfolio selection criteria ▪ Expand joint strategic planning with industry ▪ Adopt industry portfolio management techniques	▪ Stimulate larger industry investment in follow-on R&D and eventual commercialization ▪ Realize economies of scope; i.e., more market applications from technology platforms (proofs of concept)
Effective research programs and supporting technical infrastructures	▪ Update and expand research infrastructure ▪ Promote R&D and commercialization efficiency ▪ Promote more and better timed technical infrastructure	▪ Expand regional technology-based clusters involving universities, government, and industry ▪ Expand and coordinate infratechnology research to support standards early in R&D cycle ▪ Promote high-tech startups, SMEs, incubators, and accelerators ▪ Implement intellectual property rights management for cooperative search	▪ Stimulate private-sector participation early in technology life cycle by firms of all sizes, enabling product diversity, competitive industry structures, and efficient system integration ▪ Achieve shorter time to market ▪ Enhance complementary asset sourcing, risk pooling, and rapid technology transfer
Modern educational infrastructure	▪ Provide highly-skilled research and manufacturing labor force	▪ Adjust college curricula (especially within innovation clusters) ▪ Support STEM students through scholarships and career promotion ▪ Update vocational, apprenticeship programs	▪ Promote broad and deep skilled labor pool ▪ Achieve flexible and rapid skill mix adjustments in response to emerging technologies
Rapid scale-up to commercial production volumes	▪ Promote advanced manufacture process technology infrastructure ▪ Capital formation	▪ Create demonstration projects ▪ Provide production scale-up assistance ▪ Provide shared production facilities	▪ Accelerated commercialization and market share growth ▪ Increased industry rates of return and value added

CHAPTER 1 ■ Selecting Policy Tools to Expand NASA's Contribution to Technology Commercialization

Figure 1.2 shows how this problem must be managed by the R&D agency. The economic impacts that a comprehensive technology commercialization strategy can be expected to eventually generate must be preceded by interim sets of impact metrics that characterize positive results in the early and middle phases of the R&D process and then during early (post-innovation) commercialization. The necessary use of sets of interim metrics is indicated for major time phases of a technology's development. These interim impact metrics provide assurance of progress for each phase and set up the rationales for support of succeeding phases. Eventually, technology commercialization is achieved, markets are developed, and the innovating industry's contribution to the economy's growth becomes substantial. The top growth curve represents the fact that the newly commercialized technology exerts a multiplier effect on national economic growth through its enhancement of other (user) industries' productivity. Thus, the aggregate direct economic benefits to the target industries generate even larger benefits for the economy as a whole.

Short-Term
- Partnership structures and strategic alliances organized
- New research facilities and instrumentation in place
- New firm formation
- Initial research objectives met/increased stock of technical knowledge

Medium-Term
- Supply-chain structure established
- New-skilled graduates produced
- Compression of R&D cycle
- New technologies achieved
- Commercialization
  - New products
  - New processes
  - Licensing

Long-Term
- Broad industry and national economic benefits
  - Return on investment
  - GDP impacts

National Economic Impact

Multiplier Effect

Benefits to Target Industries

Year of Initial Commercialization

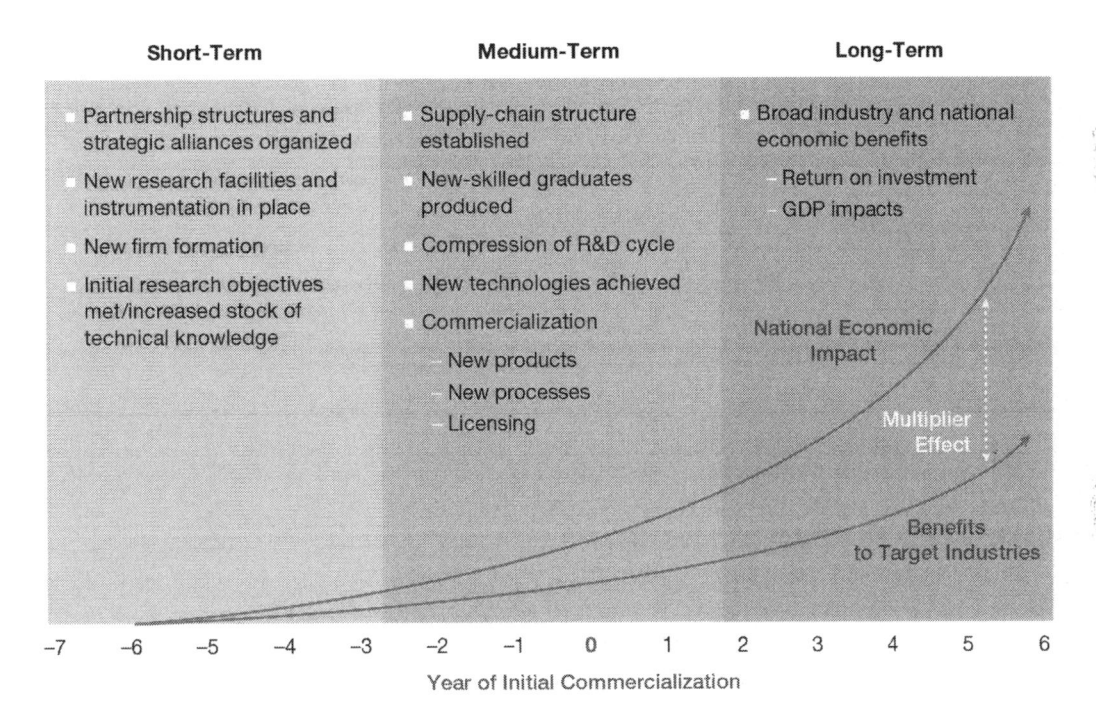

FIGURE 1.2: Nature and Timeline of Economic Impacts from Enhanced Technology Commercialization Strategies

Conclusion

NASA faces three alternative strategic directions for promoting technology commercialization:

1. Fund a limited set of largely independent research projects in LEO. This approach will provide unique but limited advances in scientific and technical knowledge with no direct path to further technology development.

2. Combine terrestrial and LEO R&D assets to fund a broader range of projects in specific technical areas, emphasizing the early pre-competitive phases of technology development (platform technologies and infratechnologies). The resulting larger scale/scope is more likely to promote follow-on investment by industry. NASA may or may not continue to support later phases of the R&D process depending on the continued existence of private underinvestment (market failures).

3. Invest in a holistic technology development infrastructure that allows joint management with industry partners of R&D project portfolios. These project portfolios will target new technology platforms and supporting infratechnologies and do so through new research infrastructures (R&D consortia, innovation clusters), followed by more limited support for the latter-phase R&D leading to eventual commercialization by industry partners.

To achieve meaningful commercialization over time, option 3 is required because modern manufacturing technologies are complex systems of hardware and software and the holistic set of funding plus institutional strategies pulls more private R&D resources and future innovators into the nascent industry (or, more accurately, set of industries; i.e., the target supply chain). It is the "system" that competes in the global marketplace. Therefore, the productivity of the entire supply chain is the critical metric in determining commercial success. This can only be achieved through a holistic approach to R&D portfolio management, combined industry-government partnerships to increase R&D efficiency, and technology transfer across component suppliers and system integrators.

NASA has unique expertise in system design and integration of the system components, having decades of experience in the design, construction and operation of systems that function in an extremely hostile environment. This expertise lies in existing NASA terrestrial and low Earth orbit laboratories and the staff that operate them. Thus, a broad implication for NASA's technology commercialization efforts is that given the likely continued focus of the ISS on the conduct of experiments associated with early-phase technology development, follow-on Earth-based efforts should be implemented to enable the entire R&D process and even initial commercialization efforts to occur until the time comes when the entire R&D process and even manufacturing will benefit from a microgravity environment.

Using the technology element model described in this chapter, NASA can adopt the appropriate policy tools for each phase of a new technology's development and effectively manage the technology's life cycle in order to achieve successful commercialization. Such "dual use" of existing and future NASA R&D assets can result in overall efficiency gains with respect to utilization of these assets and a significant contribution to U.S. efforts to meet the growing competition from the rapidly evolving technology-based global economy.

CHAPTER 2

Protein Crystallization for Drug Development

A Prospective Empirical Appraisal of Economic Effects of ISS Microgravity

Nicholas S. Vonortas[1]

Executive Summary

One of the basic missions of NASA is to use the International Space Station (ISS) to facilitate the growth of a commercial marketplace in low Earth orbit (LEO) for scientific research, technology development, observation and communications, and human and cargo transportation. While the private sector has shown some interest, and while there exists significant potential commercial applications, the future of LEO commercialization nonetheless has significant technical, financial, and policy barriers. Oft cited barriers include (1) transportation costs, frequency, and risk for cargo and research crew, (2) intellectual property rights, and (3) lack of appropriate investment and tax incentives to entice the private sector. An additional barrier has been the relative newness of the operation and, consequently, the lack of awareness of the possibilities across industry until recently.[2]

---

1 Center for International Science and Technology Policy and Department of Economics, The George Washington University.

2 Link and Maskin's chapter in this collection drives home the idea that the lack of information about previous projects has hindered the willingness of additional users to engage in R&D on the ISS. This message is picked up by the Lerner, Leamon, and Speen's chapter in this collection who provide a lot of detail through their VC interviews and literature review about what are the weak points of LEO tech business in the eyes of the private sector, what information is missing, and what types of information, regulation, and regular service provision will make business calculations more favorable toward LEO.

Disclaimer: The views and opinions of the authors do not necessarily state or reflect those of the U.S. Government or NASA.

Bioscience has long been promoted as one of the more promising applications in the effort to commercialize LEO. A specific application—expanding research in protein crystallization on the ISS's microgravity environment—could prove important in the development and design of drugs to treat diseases such as arthritis, cardiovascular disease, multiple sclerosis, osteoporosis, cystic fibrosis, and even cancer. Identifying a method for evaluating the economic additionality of microgravity for this type of application is the focus of this chapter.

Humans contain over 100,000 proteins that are vital in their everyday functions. Full understanding of protein function requires information on the three-dimensional structure of these proteins and leads to the development of better drugs that target these proteins more effectively. To obtain this information, scientists model proteins with a process called X-ray crystallography, whose effectiveness, in turn, depends on the use of good quality protein crystals.

The ISS has for long been argued as the ideal place for protein crystallization. Crystals grown in microgravity do not face the same convection forces seen on Earth—such as wind and gravity—which adversely affect the orientation and size of crystals. While crystals grow much slower in microgravity, they can also grow larger and provide massively more data points as crystals grown on Earth. Because of this, it is argued that the ISS has the potential of becoming the new frontier of pharmaceutical research, assisting in the discovery of new uses of proteins never before thought possible when tested on Earth.

The promise of opportunity notwithstanding, several constraints remain for the establishment of high-budget (i.e., viable) protein crystallization research on the ISS: proof of concept, experiment duration, size restraints, and transportation. At present, it is questioned whether there is strong enough evidence that crystal quality improvement is high enough to justify the effort of sending proteins to space (McPherson & DeLucas 2015).

There are four stages of research with heavy contribution by protein crystallization to drug development:

1. production of high quality crystals
2. collection of generic information to identify and validate possible drugs
3. modeling of a drug to a specific target during the preclinical stage
4. fine tuning of the drug during the clinical stage

The first two stages can be described as generic research, with the results being of interest across the pharmaceutical industry (also including biotech). The last two stages are appropriately described as private research of interest to individual companies. The ISS's contribution as currently envisioned falls squarely in the very first stage but its effects are very much felt all the way across. Assuming the distinction between more generic and more applied research stages holds, the policy approaches to address them are different.

On the one hand, *the more generic research stages may be addressed through a collaborative agreement between the main stakeholders on the supply side.* This is a tool well-tried out in the previous two to three decades in various sectors to alleviate the conditions for suboptimal investment in generic (or precompetitive) research by the private sector (1) due to the presence of uncertainty regarding research paths and the prospective applicability of the results within the confines of individual companies and (2) due to expectations of imperfect appropriability of the results. Compounded with the four barriers listed earlier in relation to the ISS, this creates a straightforward need for government intervention through the sanctioning of collaborative research as well as its subsidization.

On the other hand, *the more applied research stages may be addressed with better information about the expected benefits and costs and risks involved in order to facilitate private sector investment.* It is these stages of research that are the major focus of the latter part of this chapter. The purpose here is to expose a specific model that can be utilized to empirically analyze the effect of protein crystallization in space to the pharmaceutical sector. To achieve this goal one needs to take two steps. First, one must evaluate the effect of improved (larger, clearer) protein crystals on drug development. Second, one must evaluate the additionality of ISS in developing such improved crystals in microgravity as compared to other channels of doing so.

It is shown that a detailed model recently developed by RTI researchers sponsored by the New York Academy of Sciences can indeed be used in carrying out the first step, i.e., evaluating the effect of improved (larger, clearer) protein crystals on drug development. The attraction of this model is that it estimates the effect of better infrastructure for research on the productivity of R&D for a specific disease. The availability of better quality protein crystals can be considered in economic terms as better infrastructure leading to increased productivity of pharmaceutical R&D. A model like the one showcased here can help assess the benefits of improved protein crystals on drug development much more precisely than has been the case until now.

For the second step we must wait for the full results of DeLucas's study. Preliminary reported results of the study find considerable additionality. Assuming this holds, the improvement in commercial drug costs would be quite significant.

Section 1. Introduction

There are several potential commercial applications for the ISS. As has been argued throughout the years by a long list of experts and NASA itself, one of the more promising applications in the effort to commercialize LEO is bioscience. In particular, it is argued that expanding research in protein crystallization on the ISS is a low hanging fruit in terms of allowing CASIS—the manager of the federal laboratory—to develop a sustainable activity by leveraging an extant bioscience sector. There are strong expectations that the space station's microgravity environment will prove important in the development and design of drugs to treat diseases such as arthritis, cardiovascular disease, multiple sclerosis, osteoporosis, cystic fibrosis, and even cancer.

This chapter explicitly addresses this issue. In a short time period we have tried to understand the added value of the ISS microgravity environment for building better quality protein crystals than is possible on the ground. The scientific implications of our study are quite broad as it involves a huge potential area of biomedical research spanning across diseases and NIH centers. In this case study, we are more interested in the economic aspects of protein crystallization on the ISS. Within a time span of 3–4 months we have tried to understand both what scientists say about it and how economists would approach the cost-benefit analysis of this activity. At this point, we think we understand to some extent the basic scientific idea and can propose a model that can guide a collection of appropriate data from industry experts in order to quantify microgravity's "value added," while taking into account the expected benefits, expected costs, and risks involved. Obviously, as any analysis of its kind, what is proposed here is based on certain assumptions and the required data for estimation will depend on past experience in pharmaceuticals (e.g., Tufts database) as well as on certain opinions from industry experts. In other words, there is some margin for error.

This chapter proceeds as follows: Section 2 discusses the topic of protein crystallization in biomedical research as well as our current understanding regarding the contribution of microgravity in developing better quality crystals. Section 3 addresses possible policy intervention, also including the introduction of a government funded consortium to diffuse risk. Section 4 outlines a specific model that has recently been developed by RTI International, which provides an interesting quantitative framework for measuring private sector costs.[3] The outcome of this section is essentially a list of needed data and data sources. Finally, Section 5 concludes.

Section 2. Protein Crystallization and Biopharmaceutical Research

2.1 Declining Research Productivity in New Drug Development

The white paper published by the Federal Drug Administration more than 10 years ago (FDA 2004)[4] openly identified a critical challenge in biomedical research: an ever faster pace of basic science discoveries are not being translated quickly into more effective, affordable, and safe medical products for patients. In economic parlance, one would describe the problem as a decrease in biomedical research productivity. The problem was identified as concentrating on an outmoded medical product development path that has become increasingly complex, inefficient and, thus, very costly, resulting in decreased numbers of both new drug and biologic applications submitted to FDA and medical device applications. In contrast, the costs of product

---

3 We are indebted here to our co-panelist, Professor Al Link, who had participated in the specific RTI project and pointed out to us two critical outputs of it.

4 And revisited more recently (FDA 2014).

development reportedly had soared over the previous decade. If the calculation of the cost of successful drug development reported in the academic literature is anything to go by, the situation has not improved since then. On the contrary, it is getting worse (DiMasi et al. 2003, 2007, 2014).

In the FDA's view, the problem was said to be that applied sciences needed for medical product development have not kept pace with the tremendous advances in the basic sciences. The discovery process had accelerated much more rapidly than the technology development process. "[N]ot enough applied scientific work has been done to create new tools to get fundamentally better answers about how the safety and effectiveness of new products can be demonstrated, in faster time frames, with more certainty, and at lower costs"(FDA 2014, p.ii). Developers were said to often use antiquated tools and concepts, resulting in high product failure during clinical trials and significant loss of time and resources. Obviously, as in any industry, producers cross-subsidize failures from successes. In an industry where the costs of drug development are already high due to extensive regulation and complicated science, antiquated drug development structure further amplifies costs. It consequently also leads to greater attention toward a few potential megaproducts.

The FDA white paper called for a new product development toolkit containing powerful new scientific and technical methods such as animal- or computer-based predictive models, biomarkers for safety and effectiveness, and new clinical evaluation techniques (p. ii). Such a toolkit would improve predictability and efficiency along the path from laboratory concept to commercial product.

While there is no silver bullet to achieve this objective, this opens up a window of opportunity for improvements in the process of protein crystallization. Better crystals can be considered as part of better research infrastructure—exactly like better biomarkers, for instance—that would contribute to decreasing risk in the preclinical phase of new drug development and allow better compounds to be differentiated from the chuff much earlier in the follow-up clinical research phases.

2.2 Use of Protein Crystals in Biomedical Research

Humans contain over 100,000 proteins that are vital in their everyday functions. Without them, our bodies could not "repair, regulate, or protect themselves" (NASA 2015). Full understanding of protein function requires information on the three-dimensional structure of these proteins and leads to the development of better drugs that target these proteins more effectively. To obtain this information scientists model proteins with a process called X-ray crystallography, whose effectiveness depends on the use of good quality protein crystals.

2.2.1 Protein Crystallization History and Process

Protein crystallization is the process by which protein molecules are formed into 3D crystals so they can be studied much more effectively under a process called X-ray crystallography. Using X-ray crystallography, scientists study the way proteins interact with other molecules, how they undergo conformational changes, and how they perform catalysis in the case of enzymes.

Protein crystallization is a 100-year-old process that has gained renown as a drug discovery tool over history. Max von Laue has been credited for the discovery of the diffraction of X-rays by crystals in 1914. Using the process of X-ray diffraction, William Henry Bragg and William Lawrence Bragg won the Nobel Prize in 1915 for analyzing crystal structures at the atomic level. It was John Bernal and his student Dorothy Hodgkin in 1934 who produced the first X-ray diffraction photograph of a digestive enzyme, pepsin, marking what many scientists consider the beginning of protein crystallography. Hodgkin went on to discover the structure of penicillin through protein crystallography, which allowed pharmaceutical companies to mass-produce the antibiotic. Herb Hauptman and Jerome Karle, who won the Nobel Prize in 1985, found a more efficient method for determining crystal structures that improved the accuracy and time of experiments.

Hauptman and Karle's improvement of protein crystallography has become an essential tool in today's drug discovery industry. More than 85 percent of known protein structures have been discovered through the process of protein crystallography (NIH 2007, 14). In this process, a pure, highly concentrated sample of a protein is combined with a variety of liquids that will eventually evaporate, resulting in the formation of a crystallized protein. The best crystals are long, three dimensional, and tightly packed with organized molecules. Since diffraction-quality crystals can be hard to produce; thousands of samples are often created for just one protein.

After creating a successful protein crystal, X-ray diffraction is performed. Using a large machine called a synchrotron, X-rays are blasted through the crystals, which are being automatically rotated, to capture the full scope of diffraction data. Given data on diffraction patterns, proteins can be accurately modeled in three dimensions.

With the resulting three-dimensional protein model, pharmaceutical companies can design novel drugs that target a particular protein or engineer an enzyme for a specific industrial process. This development process is known as *structure-based drug design*. There are several empirical examples of protein crystallography's contribution to medicine throughout history. Notably in the 1980s, protein crystallography was vital in producing treatments for HIV. Scientists were able to model the structure of HIV protease, a protein that causes HIV to spread throughout the body. Using the three-dimensional structure, scientists engineered protein inhibitors—such as the drug nelfinavir mesylaty (Viracept)—to slow down the progress of the disease.

Pharmaceutical companies often utilize protein crystallization during the early phases of the drug discovery process, even prior to the preclinical phase. When scientists want more information about the receptor site of the target drug, protein crystallization is used to identify and validate the target. Once there is a general idea of a leading target, protein crystallization can be used further during the drug design phase to model a drug specific to the target. Specifically, scientists use software to test fit a drug candidate to the molecule's receptor site. Protein crystallization is more commonly used in the drug discovery process due to the time it takes to effectively analyze crystals for new targets.

While the key to accurate 3D protein modeling is high-quality protein crystals, protein crystallization has historically been the most difficult part of the crystallography process. Protein crystals are very fragile and can be affected by small changes in heat or pressure. With a very low margin of error, scientists must increase the sample sizes of possible crystals, this way increasing both the cost and time of experiment. Another issue is that protein structure in a crystal is not always the same as in an actual cell. Biological structures are difficult to measure solely through a representative crystal. To address this, scientists complement their analysis of the crystal structure with a protein's activity, which provides more accurate data but increases the time of experiment further. In the past 15 years, technology and automation of the process has significantly lowered costs and increased the purity of tested proteins (Netterwald 2007).[5]

In 1992, Dr. Lawrence J. DeLucas argued that the microgravity environment on the ISS would be ideal for growing better quality protein crystals. The question of the benefit from space crystallization still remains today. Why are crystals grown on the ISS better than the alternatives obtained on the ground? How much better are they and at what additional cost?

2.2.2 ISS Additionality for Protein Crystallization

The ISS is argued as the ideal place for protein crystallization due to the Space Station's microgravity environment and the existing MERLIN hardware. Crystals grown in microgravity do not face the same convection forces seen on Earth—such as wind and gravity—which adversely affect the orientation and size of crystals. While crystals grow much slower in microgravity, they provide approximately twice as many data points as crystals grown on Earth (Pool 1989). Because of this, it is argued that the ISS has the potential of becoming a frontier of pharmaceutical research, assisting in the discovery of new uses of proteins never before thought possible when tested on Earth.

The promise of opportunity notwithstanding, several constraints remain for the establishment of high-budget (i.e., viable) protein crystallization research on the ISS: proof of concept, experiment duration, size restraints, and transportation.

- **Proof of concept:** While it is generally understood that proteins crystallize better in microgravity than Earth, it is still unknown to what extent exactly this is true and which proteins may be exceptions. According to CASIS, protein resolution in low Earth orbit improves about 20–30% over crystals grown on Earth (CASIS Opportunity Map 2012, 27).

---

5 See **Appendix A** for more detailed Protein Crystallization costs on Earth surface.

- **Experiment duration:** Length of experiment may be a most important issue. What might take 1–2 weeks on Earth could take 6 months on the ISS. CASIS interviews indicate that the biotech industry would ideally like their results in 4–8 weeks.

- **Lab size:** The small volume of the ISS available for new facilities could hinder pharmaceutical companies from testing large samples and executing a scalable model.

- **Transportation to and from the ISS:** The absence of regular and frequent flights with reliable long-term schedules places a limit on what can be done and at what cost (including operational and insurance).

At present, some within the community of possible business investors argue that there is not yet strong enough evidence that crystal quality improvement is high enough to justify the effort of sending proteins to space (CASIS Opportunity Map 2012, 27). Still, CASIS finds that protein crystallization has the potential to become one of the strongest commercial applications to the ISS.

Dr. Lawrence DeLucas from the University of Alabama is currently leading a blind study of 2,000 membrane protein crystals grown in space versus Earth, which will likely be completed in the summer of 2015.[6] DeLucas' April 2014 expedition[7] purported to accomplish two goals: (1) inform scientists on the structures of membrane proteins and (2) conclusively measure the overall impact of microgravity on protein crystallization. The entire analysis is being performed as a "double blind" experiment to eliminate any perceived bias. They use a statistically relevant number of different proteins and for each analyzed protein a statistically relevant number of crystals. Assuming results in well-defined statistical confidence intervals, DeLucas' project becomes of critical importance in determining the additionality of the ISS in protein crystallization and thus enabling cost-benefit analysis of this line of activity on the ISS.[8]

The most relevant results of DeLucas' project to the question under investigation in our study will be the measured differences of protein crystals in microgravity and on Earth. Evidence of significant additional value of microgravity will support NASA's goal to commercialize this capability of the ISS and LEO more generally.

---

6 Membrane proteins make up a 67% of commercial drugs, yet information about the structure of these proteins is lacking due to inability to grow proper crystals on Earth. For this reason, Dr. DeLucas believes that membrane protein crystallization has the potential to be a strong piece of a commercial space industry. Other useful proteins with potential commercial application are "high-value aqueous proteins and protein complexes" (NASA 2015).

7 The protein samples were launched to the ISS on April 18, 2014, and returned to the investigator on October 27, 2014.

8 Participants included government labs (Oak Ridge, Los Alamos, Scripps, NIH), industry (Emerald Biostructures, Astra-Zeneca, Ixpress Genes, St. Jude Research Hospital), and 24 universities (including the University of California, the California Institute of Technology, New York University, Columbia University, University of Leeds, and Martin Luther University).

DeLucas' earlier press release notes that even a small improvement in crystals would have "a significant impact on scientists' ability to use the resulting structures to provide insights into biological mechanisms" (NASA 2015). This "significant impact" can be critical in reducing the cost of existing infrastructure and R&D.

At the very last moment of this writing, Larry DeLucas announced preliminary results of his study (DeLucas 2015). The data were interesting and the pictures compelling in the sense of pointing out that in certain cases of those examined the obtained protein crystals were largely improved in microgravity as compared to those created on Earth. More information regarding the costs of and demand for protein crystallization in microgravity can be seen in **Appendix A** (p. 42).

Section 3. Policy Considerations

One can distinguish four stages of research with heavy contribution by protein crystallization to drug development that may be liable for possible policy intervention. These stages are:

1. production of high quality crystals
2. collection of generic information to identify and validate possible drugs
3. modeling of a drug to a specific target during the preclinical stage
4. fine tuning of the drug during the clinical stage

The first two stages can be described as generic research, with the results being of interest across the pharmaceutical industry (also including biotech). The last two stages are appropriately described as private research of interest to individual companies. The ISS's contribution as envisioned comes squarely in the very first stage but its effects are very much felt all the way across. Assuming the distinction between more generic and more applied research stages holds, the policy approaches to address them are different. The more generic stages may be addressed through a collaborative agreement between the main stakeholders on the supply side. The more applied stages may be addressed with better information about the expected benefits and costs and risks involved in order to facilitate private sector investment.

3.1 Collaborative Agreement

There has been a long strand of research in the economics, business management, and policy literatures on collaborative research.[9] The reason for the development of this extensive literature in its earlier phases in the 1980s and 1990s sounds tantalizingly

---

9 Starting in the early 1980s, the literature on collaborative R&D has grown really large. There are several surveys of this literature, some contributed by two members of this panel. See, for example, Vonortas (1997), Jankowski, Link, and Vonortas (2001), Vonortas and Zirulia (2015), and Hagedoorn, Link and Vonortas (2000).

similar to the first two stages of protein crystallization contribution seen above. In a few words, generic (or precompetitive) research creates the conditions of suboptimal investment by the private sector (1) due to the presence of uncertainty regarding research paths and the prospective applicability of the results within the confines of individual companies and (2) due to expectations of imperfect appropriability of the results (Arrow 1962, Nelson 1959). Compounded with the four barriers listed earlier in relation to the International Space Station (Section 2.2.2), this creates a straightforward need for government intervention through the sanctioning of collaborative research as well as its subsidization.

3.2 Key Questions

As the manager of the U. S. National Lab, CASIS already subsidizes research related to protein crystallization on the ISS. The question, of course, is if the subsidy is enough and if it is used to support applied research rather than just basic research.[10]

In order to understand whether the subsidy is at the appropriate level, one needs to consider several issues:

1. Is there additionality of the ISS microgravity environment in building better protein crystals?
2. Assuming significant additionality, should there be a collaborative undertaking involving protein crystallization using the ISS microgravity environment?
3. What does it take to build such a collaborative agreement? Does it make sense to build it solely among American firms and research institutes or, given that ISS is a 15-nation endeavor, build it across all partners?
4. How should the latter two stages of protein crystallization's contribution to the private sector—specifically pharmaceutical research—be considered in these calculations?
5. What are the true private and social benefits to consider in a proper cost-benefit analysis?

We must await the full results of DeLucas' ongoing study (Section 2.2.2) to gain better insight into the first question. Assuming significant additionality, the argument to answer the second question should be affirmative. A recent study of the National Research Council (2015) summarizes this argument for collaborative generic research in a different technology area (flexible electronics) in a form that can readily be applied here. But the argument has been settled long ago in detailed investigations of the theoretical and empirical aspects of the rationale of cooperative R&D needed

---

10 Basic research produces, of course, a public good that is supported by the public purse.

to support legislation like the National Cooperative Research Act (NCRA) of 1984 and its sequel, the National Cooperative Research and Production Act (NCRPA) of 1993.[11]

Calculating the optimal level of support for this type of consortium (third question) could be the subject of an entirely separate analysis. The government could take several approaches in determining the size of public expenditure. One approach is to match private investment, as seen in other consortiums. Alternatively, the government could fund all costs associated with obtaining better protein crystals from space (e.g., space travel and samples). Generic protein crystals could then be distributed (or sold) to pharmaceutical companies for further collaboration and examination. No matter what method, the government must contribute enough to ease the uncertainty and risk involved from obtaining space crystals.

The remaining two questions pertain to the cost-benefit analysis of protein crystallization contribution to new drug development. It would be easier, but misguided, to build the argument on the basis of the first two stages of protein crystallization contribution only (**first paragraph Section 3**). All four stages should be considered instead. The estimation can follow the standard approach in health economics of calculating the costs and benefits of new drug development, then trying to disentangle the purely private from the purely social benefits.

Section 4. Modeling Drug Development Costs

The purpose of this section is to explore a specific model that can be utilized to empirically analyze the effect of protein crystallization in space (ISS) to the pharmaceutical sector. To achieve this goal one needs to take two steps. First, one must evaluate the effect of improved (larger, clearer) protein crystals on drug development. Second, one must evaluate the additionality of ISS in developing such improved crystals in microgravity as compared to other channels of doing so.

4.1 Phases of Pharmaceutical Research

The literature has advanced an aggregate model paradigm that depicts the drug development process as pretty much linear with several phases.

Preclinical. The preclinical phase of the drug development process typically encompasses discovery and preclinical development testing. *Discovery programs* aim at synthesizing compounds that then undergo *preclinical testing* in assays and animal models.

---

11 Vonortas (1997) provides an almost exhaustive review of the theoretical and empirical arguments utilizing mainstream economic concepts from transaction cost economics, public goods and externalities, and investment behavior under conditions of uncertainty, impactedness, and opportunism. Hemphill and Vonortas (2003) expand to arguments from the management literature such as real options and competitive advantage.

Clinical. The clinical part of the drug development process refers to human testing. Clinical testing typically proceeds through three successive phases:

- **Phase I:** a small number of usually healthy volunteers are tested to establish safe dosages and to gather information on the absorption, distribution, metabolic effects, excretion, and toxicity of the compound.

- **Phase II:** trials are conducted with human subjects who have the targeted disease or condition. These trials are conducted on larger numbers of subjects than in phase I (maybe hundreds) and are designed to obtain evidence on safety and preliminary data on efficacy.

- **Phase III:** testing typically consists of a number of large-scale trials designed to establish efficacy and to uncover side effects that occur infrequently. The number of subjects is now the largest and can total in the thousands.

4.2 Effect of Improved Protein Crystals: Model Development and an Example[12]

We use as a base the model by Scott et al. (2014) [also in New York Academy of Sciences (2013)], which measures the cost of drug development as a function of cost, time, and risk. The expected cost of developing a new drug is given by the sum of the risk-adjusted, capitalized cost of each phase of development:

$$\left(c \int_{t_{end}}^{t_{start}} e^{rt/12}\, dt\right)/p = \left(\frac{c}{p}\right)\left(\frac{12}{r}\right)(e^{rt_{start}/12} - e^{rt_{end}/12})$$

where t_{start} denotes time in months from start of phase to date of new drug approval
t_{end} denotes time in months from end of phase to date of new drug approval
c is cost per month per compound in phase
p is the probability that a compound undergoing this phase of development is ultimately approved for marketing
r is cost of capital as an annual interest rate.

Drug development is a lengthy process, implying substantial time costs to investing in R&D long before any potential returns can be earned. The time costs of drug development can be captured by capitalizing costs forward to the point of marketing approval at an appropriate discount rate. Capitalization is achieved by continuous compounding at the annual interest rate r, which can be set at 10.5–11% through the

---

12 This section borrows heavily from Scott et al. (2014) and New York Academy of Sciences (2013). It also consults extensively DiMasi and Grabowski (2007) and DiMasi, Hansen, and Grabowski (2003). A new study recently released by the Tufts Center for Drug Development (DiMasi 2014) was not available to us at the time of this writing.

CAPM model or otherwise for the biopharmaceutical industry (Harrington 2012; DiMasi and Grabowski 2007).

Capitalized costs are the sum of out-of-pocket (money) costs and time costs. It should be noticed that in order to obtain time costs one needs—in addition to an appropriate discount rate—a timeline over which out-of-pocket costs are capitalized forward to marketing approval. Using data for several compounds, DiMasi and Grabowski (2007) and DiMasi et al. (2003) estimate average phase and regulatory review lengths. On the other hand, in their case study on Alzheimer's disease Scott et al. (2014) obtain estimates of duration—as well as estimates of cost and probability (to progress to next phase)—by experts in Alzheimer's research and drug development. They used data provided by experts. See Tables 2.1, 2.2, and 2.3.

TABLE 2.1: Duration of Drug Development Phases (Months)

PHASE	TYPICAL NEW BIOPHARMACEUTICAL (DiMasi and Grabowski 2007)	ALZHEIMER'S DISEASE* EXISTING (Scott et al. 2014)	ALZHEIMER'S DISEASE* IMPROVED
Preclinical	52.0	50.1	49.9
Phase I	12.3	12.8	12.6
Phase II	26.0	27.7	25.2
Phase III	33.8	50.9	39.4
Regulatory Review	18.2	18.0	16.9

* Durations are presented with 95% confidence
Source: Adapted from New York Academy of Sciences (2013), Table B-2.

TABLE 2.2: Transition Probabilities* Between Phases

PHASE	TYPICAL NEW BIOPHARMACEUTICAL (DiMasi and Grabowski 2007)	ALZHEIMER'S DISEASE* EXISTING (Scott et al. 2014)	ALZHEIMER'S DISEASE* IMPROVED
Phase I to II (1)	0.71	0.67	0.69
Phase II to III (2)	0.44	0.47	0.42
Phase III to Approval (3)	0.68	0.24	0.58
Phase II to Approval (2) × (3)	0.30	0.11	0.24
Phase I to Approval (1) × (2) × (3)	0.21	0.07	0.16
Ratio of Phase II failures to total failures in Phase II and III	0.80	0.60	0.77

* Average reported
Source: Adapted from New York Academy of Sciences (2013), Table B-3.

TABLE 2.3: Average Costs (typical new biopharmaceutical—$M)

PHASE	OUT-OF-POCKET (MONTHLY) TYPICAL NEW BIOPHARMACEUTICAL ($M/ MOLECULE IN DEVELOPMENT) (DiMasi and Grabowski 2007; DiMasi et al. 2003)	CAPITALIZED AT 11% TYPICAL NEW BIOPHARMACEUTICAL ($M/NEW DRUG APPROVED) (DiMasi and Grabowski, 2007; DiMasi et al. 2003)
Preclinical	0.72	510
Phase I	2.73	338
Phase II	2.00	312
Phase III	5.64	385
Total	**11.09**	**1,565**

Source: Adapted from New York Academy of Sciences (2013), Table B-4.

The older estimates above have been calculated on the basis of data from the Tufts database on drugs developed by traditional pharmaceutical companies (DiMasi et al. 2003) and by biotechnology companies (DiMasi and Grabowski 2007). The latter paper concentrated on the types of molecules on which biotech companies focus, specifically recombinant proteins and monoclonal antibodies (mAbs). Thirteen of a total seventeen examined compounds[13] first entered clinical testing during 1990–2003; the remaining four compounds examined in this study were from the Tufts database. The newer estimates specific to Alzheimer's disease were developed on the basis of detailed expert interviews from the pharmaceutical industry and academia (Scott et al. 2014; New York Academy of Sciences 2013).

Tables 2.4 and 2.5 use steps mathematically identical to the formula presented earlier in this section to highlight (1) the expected cost of entering a drug candidate in Phase I trials and (2) the total capitalized cost of a new drug approval. The important takeaway here is how even small (apparent) changes in the cost, likelihood of successful completion, and time length of a phase lead to dramatic decreases in overall costs of successful drug development.

Finally, Table 2.6 shows the estimated costs of developing a disease-modifying drug for Alzheimer's across the industry (New York Academy of Sciences 2013). This is the typical way the literature has reported estimates for drug development costs and include the cost of failures by multiple companies expected by the interviewed experts before one drug is approved by the FDA for marketing. Notice that totals of each column match the totals in Tables 2.4 and 2.5 respectively.

---

13 The sample consisted of 9 recombinant proteins and 8 mAbs.

CHAPTER 2 ▪ Protein Crystallization for Drug Development

TABLE 2.4: Cost of Alzheimer's Disease-Modifying Development with Existing Infrastructure

EVENTUAL OUTCOME FOR A COMPOUND ENTERING PHASE I	OUT-OF-POCKET COST ($M)	COST CAPITALIZED TO DATE DEVELOPMENT STOPS OR DRUG APPROVED ($M)	COST AT PHASE I START (PRESENT VALUE) ($M, 11% DISCOUNT RATE)	PROBABILITY
Development stops after Phase I	71	89	79	0.33
Development stops after Phase II	126	177	122	0.35
Development stops after Phase III	413	648	280	0.24
Drug is approved	413	765	280	0.07

Expected present-value cost = (79 × 0.33) + (122 × 0.35) + (280 × 0.24) + (280 × 0.07) = $157M
Cost per new drug approval = $157M / 0.07 = $2,087M
Cost capitalized to date of drug approval = $2,087M × $e^{(109.4)(0.11/12)}$ = $5,693M
(Phase I starts an average of 109.4 months prior to approval)

Notes
1. Numbers have been rounded. For example, $2,087M comes from dividing approximately $156.5M by approximately 0.075.
2. Out-of-pocket cost is the monthly cost for each phase (Table 2.3) times the number of months spent in that phase (Table 2.1): 71 = (0.72)(50.1) + (2.73)(12.8); 126 = 71 + (2.00)(27.7); 413 = 126 + (5.64)(50.9).
3. Present-value cost is the value of costs incurred at the beginning of Phase I.
4. Probabilities are derived from Table 2.2 (they may not sum to 1 because of rounding): 0.33 = 1 − 0.67, 0.35 = (0.67)(1 − 0.47), 0.24 = (0.67)(0.47)(1 − 0.24).

Source: New York Academy of Sciences (2013), Table B-5.

TABLE 2.5: Cost of Alzheimer's Disease-Modifying Development with Recommended Infrastructure

EVENTUAL OUTCOME FOR A COMPOUND ENTERING PHASE I	OUT-OF-POCKET COST ($M)	COST CAPITALIZED TO DATE DEVELOPMENT STOPS OR DRUG APPROVED ($M)	COST AT PHASE I START (PRESENT VALUE) ($M, 11% DISCOUNT RATE)	PROBABILITY
Development stops after Phase I	70	87	78	0.31
Development stops after Phase II	121	167	118	0.40
Development stops after Phase III	343	507	250	0.12
Drug is approved	343	592	250	0.17

Expected present-value cost = (78 × 0.31) + (118 × 0.40) + (250 × 0.12) + (250 × 0.17) = $144M
Cost per new drug approval = $144M / 0.17 = $855M
Cost capitalized to date of drug approval = $855M × $e^{(94.1)(0.11/12)}$ = $2,027M
(Phase I starts an average of 94.1 months prior to approval)

Notes
1. Numbers have been rounded. For example, $855M comes from dividing approximately $143.5M by approximately 0.168.
2. Out-of-pocket cost is the monthly cost for each phase (Table 2.3) times the number of months spent in that phase (Table 2.1): 70 = (0.72)(49.9) + (2.73)(12.6); 121 = 70 + (2.00)(25.2); 343 = 121 + (5.64)(39.4).
3. Present-value cost is the value of costs incurred at the beginning of Phase I.
4. Probabilities are derived from Table 2.2 (they may not sum to 1 because of rounding): 0.31 = 1 − 0.69, 0.40 = (0.69)(1 − 0.42), 0.12 = (0.69)(0.42)(1 − 0.58).

Source: New York Academy of Sciences (2013), Table B-6.

TABLE 2.6: Average Costs (Alzheimer's disease—$M)

PHASE	CAPITALIZED AT 11% EXISTING INFRASTRUCTURE ($M/NEW DRUG APPROVED)	CAPITALIZED AT 11% RECOMMENDED INFRASTRUCTURE ($M/NEW DRUG APPROVED)
Preclinical	1,658	642
Phase I	1,193	458
Phase II	1,048	387
Phase III	1,794	539
Total	**5,693**	**2,027**

Source: Adapted from New York Academy of Sciences (2013), Table B-4.

4.3 Summing Up

The same, or very similar, model to that outlined in **Section 4.2** can be utilized to assess the effect of improved protein crystals on R&D targeting a specific disease. Assuming the use of average data regarding the effects across various types of diseases—across all types of cancer or all types of diabetes, for example—one can estimate average estimates for the industry. The important limitation currently is missing data on

- the duration (number of months) of the various phases of drug development,
- the cost per month, and
- the probabilities that the compounds under investigation will pass through each phase successfully.

Appendix 1 of the New York Academy of Sciences (2013) study provides the questionnaire utilized in that study[14] to elicit such data through interviews with a significant number experts including 27 industry representatives and 5 academics. The majority of the industry interviewees were reported to be at the level of vice president (or equivalent) and above in 11 companies pursuing Alzheimer's disease drug discovery and development, and they themselves were responsible for research either on Alzheimer's disease directly or on diseases of the central nervous system more broadly. The five non-industry interviewees each had more than 20 years experience in Alzheimer's related research.

The interviewees were provided with estimates on cost, cycle times, and transition possibilities found for pharmaceuticals in general and were asked to customize them for Alzheimer's disease twice: with and without a new environment of better infrastructure which was clearly defined.

---

14 Troy Scott, Alan O'Connor, and Al Link for RTI and Diana L. van de Hoef for the New York Academy of Sciences.

How does this translate to our research project on protein crystallization in microgravity? The availability of better quality protein crystals can be considered in economic terms as better infrastructure leading to increased productivity of pharmaceutical R&D.[15] A model like the one shown in this section should then help assess the benefits of improved protein crystals on drug development much more precisely than has been the case until now.

If a questionnaire is to be utilized to collect data, a well-defined target must be selected in order for the interviewed experts to provide more accurate answers. A follow-up final step would be to calculate the societal cost savings from reducing disease.

Section 5. Concluding Remarks

Protein crystallization is an essential part of protein crystallography—the main process used in drug discovery today. Advancements in the technology of protein crystallization over the last hundred years have been remarkable. The prospect of improving protein crystals through the use of the International Space Station has the potential to be the next great chapter in drug discovery history.

For the private sector, better protein crystals mean better 3D models of proteins in the preclinical stage. The improved modeling should carry over in clinical stages as well, advancing overall drug infrastructure. Recent relevant work supported by the New York Academy of Sciences for a specific disease strongly indicates that even marginal improvements in infrastructure can *significantly* reduce costs in overall drug development.

The leap from Earth crystals to drastically pricier (yet better quality) space crystals may be too risky of an investment to expect from private pharmaceutical companies, even with significant projected cost reductions. A government-subsidized consortium may be the best plan to alleviate some of the early-stage, precompetitive financial risk. The government therefore may wish to fund space missions for protein crystals for generic use, where improvements in these proteins' 3D models could lead to extensive efficiencies in drug discovery and development for a wide range of diseases. While pharmaceutical companies would still be in competition, improving their overall infrastructure of basic protein knowledge could provide strong public and private benefit.

While private cost reduction can be measured with the model seen in **Section 4**, public benefit from disease treatment and potential government revenue are difficult to measure. The potential costs to pharmaceutical companies—even with the easing of a consortium—must also be measured to gauge private interest. If public/private benefits outweigh public/private costs, then investment in protein crystallization on space is a fiscally viable use of ISS resources.

---

15 One could think of it as analogous to what economists argue about basic research and its impact on applied research and development.

References

Arrow, Kenneth (1962). "Economic welfare and the allocation of resources for invention," in Richard R. Nelson (ed.) *The Rate and Direction of Inventive Activity*, Princeton University Press.

CASIS Opportunity Map (2012). "Maximizing the Value of the CASIS platform—Bioscience Opportunity Map."

DeLucas Larry (2015). "Comprehensive Analysis of Microgravity Protein Crystallization", Presentation, March 19.

DiMasi, Joseph A., Ronald W. Hansen, and Henry G. Grabowski (2014). "Innovation in the Pharmaceutical Industry: New Estimates of R&D Costs," Report, Tufts Center for the Study of Drug Development, Tufts University. *http://csdd.tufts.edu/files/uploads/Tufts_CSDD_briefing_on_RD_cost_study_-_Nov_18,_2014.pdf* and *http://csdd.tufts.edu/news/complete_story/cost_study_press_event_webcast*.

DiMasi, Joseph A. and Henry G. Grabowski (2007). "The cost of biopharmaceutical R&D: Is biotech different?" *Managerial and Decision Economics*, 28: 469–479.

DiMasi, Joseph A., Ronald W. Hansen, and Henry G. Grabowski (2003). "The price of innovation: New estimates of drug development costs," *Journal of Health Economics*, 22: 151–185.

Food and Drug Administration (2004). "Innovation Stagnation: Challenge and Opportunity on the Critical Path to New Medical Products", US Department of Health and Human Services. Revised 2014.

Hagedoorn, John, Albert N. Link, and Nicholas S. Vonortas (2000). "Research Partnerships," *Research Policy*, 29(4–5): 567–586.

Harrington, S. E. (2012). "Cost of Capital for Pharmaceutical, Biotechnology, and Medical Device Firms" In P. M. Danzon and S. Nicholson (eds), *The Oxford Handbook of the Economics of the Biopharmaceuticals Industry* (pp. 75–99). Oxford University Press.

Hemphill, Thomas and Nicholas S. Vonortas (2003). "Strategic Research Partnerships: A Managerial Perspective," *Technology Analysis and Strategic Management*, 15(2): 255–271.

Jankowski, John E., Albert N. Link, and Nicholas S. Vonortas (eds) (2001). *Strategic Research Partnerships*, National Science Foundation.

McPherson, Alexander and DeLucas, Larry (2015). "Microgravity Protein Crystallization" *NPJ Microgravity*. 1, 15010. Doi:10.1038/npjmgrav.2015.10, published online 3 September 2015.

NASA (2015) "Commercial Protein Crystal Growth—High Density Protein Crystal Growth Modified (CPCG-HM)," U.S. Department of Defense.

National Institutes of Health (2007). "The Structures of Life," U.S. Department of Health and Human Services.

National Research Council (2015). *The Flexible Electronics Opportunity*, Board on Science, Technology and Economic Policy, Washington, DC: National Academies Press.

Nelson, Richard R. (1959). "The simple economics of basic scientific research," *Journal of Political Economy*, June: 297–306.

Netterwald, James, ed. (2007). "Crystallography Illuminates Drug Targets." *Drug Discovery & Development*. N.p., 6 Sept. Web. 09 Apr. 2015.

New York Academy of Sciences (2013). "Economic Analysis of Opportunities to Accelerate Alzheimer's Disease Research and Development," Report, The New York Academy of Sciences.

Pool, Robert. (1989). "Zero Gravity Produces Weighty Improvements," *Science* 246.4930: 580.

Scott, Troy J., Alan C. O'Connor, Albert N. Link, Travis J. Beaulieu (2014). "Economic analysis of opportunities to accelerate Alzheimer's disease research and development", *Annals of the New York Academy of Sciences,* 1313: 17–34.

US Department of Defense (1997). "SEMATECH 1987–1997: A Final Report to the Department of Defense," U.S. Department of Defense.

Vonortas, Nicholas S. (1997). *Cooperation in Research and Development*, Kluwer Academic Publishers.

Vonortas, Nicholas S. and Lorenzo Zirulia (2015). "Strategic Technology Alliances and Networks," *Economics of Innovation and New Technology*, 24(5): 490–509.

Appendix A. New Details Regarding Costs of and Demand for Protein Crystallization in Microgravity[16]

Estimated Protein Crystallization Costs on Earth[17]

PROTEIN TYPE	DESCRIPTION	COST
Aqueous protein	Most common	$10K–$30K
Large/complex proteins	Make up a substantial percentage of proteins of interest to pharmaceutical companies	$100K
Membrane proteins, protein-protein complexes, protein-ligand complexes	Wide range of applications	$1M

Added Protein Crystallizations Costs in Space (Estimations)[18]

ISS SPECIFICATION	COST
Price/lb in flight	$15K/lb
Flight cost for protein crystallization unit	$1.5M

Market for Protein Crystallization[19]

PROJECT PERIOD	NUMBER OF CENTERS	TOTAL COSTS	NUMBER OF PSI STRUCTURES	AVERAGE VALUE OF PROTEIN STRUCTURE
July 2005–June 2010	14	$325M (funded mostly by National Institute of General Medical Sciences)	4,800	$325M/4,800 = $67K

There is large demand for high quality protein crystals. There are over 8,000 highly valued proteins that have not been crystallized with significant quality on Earth and over 100,000 of general interest to medicine, according to the NIH.[20] As a response, the NIH created the Protein Structure Initiative (PSI) to assemble a collection of three-dimensional protein structures for research and drug development.

---

16 All information courtesy of Lynn Harper (June 8 e-mail). It is preliminary; strong caveats apply.

17 Cost includes isolation, crystallization, diffraction, and determination of protein structure.

18 One ISS protein crystallization unit can hold about 1000 crystallization samples (25 lb) plus the incubator holding the proteins (80 lb).

19 PSI data as of August 2010.

20 See *http://www.nature.com/nrd/journal/v5/n10/full/nrd2132.html* for list generated in 2006.

CHAPTER 3

Does Information About Previous Projects Promote R&D on the International Space Station?

Albert N. Link[1]
Eric S. Maskin[2]

> America has always been greatest when we dared to be great. We can reach for greatness again. We can follow our dreams to distant stars, living and working in space for peaceful, economic, and scientific gain. Tonight, I am directing NASA to develop a permanently manned space station and to do it within a decade.
>
> —President Ronald Reagan
> State of the Union Address January 25, 1984

Section 1. Introduction

The epigram above marks 1984 as the beginning of U.S. efforts to develop a space station.[3] Later in that decade, NASA succeeded in creating a partnership with Canada, Japan, and several European nations, and a formal agreement was reached to build Space Station Freedom in 1988 (Smith 2001). Following the U.S.–Russian cooperation agreement established by President George H.W. Bush, President Bill Clinton

---

1 University of North Carolina at Greensboro.
2 Harvard University.
3 According to Smith (2001, p. 1): "NASA had wanted permission to build a space station that could be permanently occupied by rotating crews since the late 1960s. Budget constraints, however, forced the agency to choose between a space station and a reusable space transportation system—the space shuttle. NASA decided to build the shuttle first. Soon after the first shuttle launch in 1981, NASA intensified efforts to win approval for a permanently occupied space station. President Reagan's 1984 speech was the culmination."

Disclaimer: The views and opinions of the authors do not necessarily state or reflect those of the U.S. Government or NASA.

announced in 1993 that Russia would join with the United States in the effort to build the International Space Station (ISS). The ISS represented the culmination of collaboration among nations for the purpose of designing, developing, operating, and utilizing a permanently occupied civil space station (NASA 2007, p. 3).

Fifteen years after President Reagan's 1984 State of the Union Address, two elements of the ISS were launched into low Earth orbit (LEO). In November 1998 the Zarya module was launched by Russia, and in December 1998 the Utility module was launched by NASA; both modules were paid for by NASA (Smith 2001). Although the ISS was not complete, its first crew launched on October 31, 2000, and docked on November 2, 2000.[4] In 2010, the ISS was completed at a cost of over $100 billion.

The National Aeronautics and Space Administration Authorization Act of 2005 (Public Law 109-155) stated in Section 501(a): "It is the policy of the United States to possess the capability for human access to space on a continuous basis." And in Section 505(a, b), the Act stated:

> It is the policy of the United States to achieve diverse and growing utilization of, and benefits from, the ISS... The ISS will... support any diagnostic human research, on-orbit characterization of molecular crystal growth, cellular research, and other research that NASA believes is necessary to conduct, but for which NASA lacks the capacity to return the materials that need to be analyzed to Earth.

Toward accomplishing these policy directives, Section 507(a, b) of the Act designated the ISS as a national laboratory (NL) and it charged the Administrator of NASA to "enter into a contract with a nongovernmental entity to operate the ISS national laboratory."

In February 2011, NASA issued a Cooperative Agreement Notice (CAN) (NASA 2011, p. 1):[5]

> The National Aeronautics and Space Administration (NASA) is soliciting proposals for competitive evaluation and award of a Cooperative Agreement to a non-profit entity to develop the capability to implement research and development (R&D) projects utilizing the International Space Station (ISS) National Laboratory (NL) and to manage the activities of the ISS NL.

The CAN defined the mission of the NL (NASA 2011, p. 4):

> The NL Entity will be responsible for maximizing the value of ISS to the Nation by developing and managing a diversified R&D portfolio based on U.S. national needs for basic and applied research and by using the ISS as a venue for Science, Technology, Engineering, and Mathematics (STEM) educational activities.

---

4 See *http://www.nasa.gov/mission_pages/station/main/onthestation/facts_and_figures.html*.

5 The solicitation document is at: *http://nspires.nasaprs.com/external/solicitations/summary.do?method=init&solId={BFE2288E-88A6-1FE1-DCFB-F94B9102C646}&path=future*.

On July 13, 2011, the Center for the Advancement of Science in Space (CASIS) was selected to manage the ISS.[6]

This chapter is part of the broader LEO Economic Development and Industrial Policy Study sponsored by NASA. Specifically, we propose a way that NASA might promote more commercial R&D conducted on the ISS, and thereby expand the commercial possibilities of LEO.[7]

ISS R&D is conducted on behalf of (1) academic institutions and research institutions (hereafter, universities) and (2) commercial entities (hereafter, firms). In this chapter we focus only on the second group, the commercial firms. Our emphasis on firms should not be interpreted as downplaying the importance of university research. Rather, this chapter represents an exploratory effort to offer strategic guidance to NASA for expanding ISS R&D for commercial purposes, and so emphasis on firms is a natural starting point. Of course, over time university-based research leads to the development of new technologies, and those technologies can be transferred to the private sector and later commercialized. However, the technology transfer process takes time, and ISS R&D is too recent to have yielded much data on this process yet.

In **Section 2** of the paper, we outline our particular proposal for expanding the commercial possibilities of LEO. Specifically, we present a model that suggests how NASA might promote more R&D on the ISS by providing candidate firms with more information about previous R&D projects and experiments. In **Section 3**, we describe some of the R&D being conducted on the ISS by universities and firms. In **Section 4**, we "test" our model with the results of an exploratory survey we conducted. The survey involved both firms that had completed R&D projects on the ISS and those that had recently applied to do such projects. Finally, in **Section 5**, we offer some recommendations to NASA about how it might implement our proposed strategy.

Section 2. A Model for Promoting R&D on the ISS

Our approach rests on the idea that if firms have more information about previous R&D projects and experiments on the ISS, then they are likely to do more R&D there themselves.

More specifically, suppose that information about other firms' R&D experiments is correlated with a firm's rate of return from current R&D on the ISS. Assume, in addition, that there are diminishing returns to the firm's investment in R&D (at least, after a certain point). Then, if NASA provides this past information to the firm,

---

6 See *http://www.iss-casis.org/About/AboutISSNationalLab.aspx* and *http://www.iss-casis.org/About/AboutCASIS.aspx*.

7 This objective follows from The White House memorandum on Accelerating Technology Transfer and Commercialization of Federal Research in Support of High-Growth Businesses. Therein: "Agencies with Federal laboratories shall develop plans that establish performance goals to increase the number and pace of effective technology transfer and commercialization activities in partnership with non-federal entities, including private firms, research organizations, and non-profit entities."

uncertainty about the firm's rate of return will be reduced, and so the firm will wish to increase its R&D investment.

Formally, the argument can be made formal as follows: Suppose that if a firm invests a dollars in R&D on the ISS, it gets back a gross return $f(\tilde{x}a)$, where f is an increasing, strictly concave function (the concavity reflects the idea that there are eventually diminishing returns to additional investment). The firm wants to maximize its expected net return and so solves

$$\max_a Ef(\tilde{x}a) - a. \qquad (1)$$

Let a^* be the value of a that solves (1).

Next, suppose that $\tilde{y}$ is a random variable correlated with $\tilde{x}$, and that the realization of $\tilde{y}$ —call this y—can be discovered by NASA (perhaps by surveying firms that have previously invested in ISS R&D). Assume that NASA has the option of committing itself to revealing the realization y to the firm. In that case, the firm solves

$$\max_a \left[Ef(\tilde{x}a) - a \,|\, y \right]. \qquad (2)$$

For each y, let $a^{**}(y)$ be the value of a that solves (2) and let $a^{**} = Ea^{**}(\tilde{y})$ It can be shown that

$$a^{**} > a^*. \qquad (3)$$

That is, the expected investment in ISS R&D is higher when NASA reveals y than when it does not.

Less formally, the reason why the firm's expected investment in ISS R&D is higher when it knows the value of y is as follows. Because $\tilde{x}$ and $\tilde{y}$ are correlated, knowledge of y decreases the variability (i.e., the dispersion) of the firm's return on investment (the dispersion of $\tilde{x}$ conditional on y is lower than $\tilde{x}$'s unconditional dispersion. Since there are diminishing returns to investment, this reduction in variability encourages the firm to invest more.

For example, suppose that if a firm invests a dollars in R&D it gets back

$$\$\sqrt{2a} \text{ with probability } \tfrac{1}{2}$$

and

$$\$\sqrt{4a} \text{ with probability } \tfrac{1}{2}$$

The firm will choose a to maximize

$$\tfrac{1}{2}\sqrt{2a} + \tfrac{1}{2}\sqrt{4a} - a,$$

i.e., $a = \tfrac{3}{8} + \tfrac{\sqrt{2}}{4}$ Now, suppose that by giving the firm information about R&D projects, NASA can resolve the uncertainty about the rate of return, i.e., either the firm knows it will get $\sqrt{2a}$ or it knows that it will get $\sqrt{4a}$. In the former case, it will choose a to maximize

$$\sqrt{2a} - a,$$

i.e., $a = \frac{1}{2}$. In the latter case, it will choose a to maximize

$$\sqrt{4a} - a,$$

i.e., $a = 1$. Thus, the average investment when NASA provides information is $\frac{3}{4}$, which is bigger than $\frac{3}{8} + \sqrt{2}/4$, the investment without information.

Section 3. Background Information on CASIS Activities

As noted in the Introduction, CASIS was selected in July 2011 to manage research and technology projects within the ISS U.S. National Laboratory. According to CASIS's Annual Report for FY 2013, its mission (p. 1), which is based on the 2011 CAN, is:

> To enable and increase the use of the International Space Station U.S. National Laboratory as a unique dynamic platform for scientific discovery, technology development and education for the benefit of life on Earth. CASIS is responsible for maximizing the value of the ISS to the nation by developing and managing a diversified R&D portfolio based on U.S. national needs for basic and applied research and by using the ISS as a venue for Science, Technology, Engineering and Mathematics (STEM) educational activities.

Its vision (p. 5) is:

> To fully realize the unique scientific, technological and educational potential of the ISS NL by focusing both outwardly—toward exposing the scientific, technological and educational communities to the benefits that can come from research and operations in space—and inwardly—toward improving humankind's wellbeing on Earth. The outward and inward-looking aspects of the CASIS Vision are intertwined and will require close collaboration with NASA, other government agencies, research and educational institutions, industry partners and commercial entities committed to exploring the intellectual, technological and economic opportunities offered by space. An important focus of the CASIS mission is to engage and connect to new stakeholders who have not been traditionally involved with NASA or with space research.

And its goals (p. 6) are to:

- Establish a robust "innovation cycle" where first-class science will drive the development of technologies, new intellectual property and commercial opportunities, which in turn drive new ideas and novel first-class science.

- Utilize the ISS for developing new capabilities based on existing proof-of-concept technologies, while allowing time for longer-term scientific commercial initiatives to develop.

- Undertake a strong public outreach promoting the value of the ISS NL to the nation, and establish the ISS NL as a leading laboratory and environment for STEM education.

Toward fulfilling its mission, vision, and goals, CASIS solicits proposals and considers unsolicited proposals for R&D projects to be conducted on the ISS. Proposals are welcome from both universities and firms.

Table 3.1 describes solicited proposals from FY 2012 through FY 2014.[8] Clearly, the majority of solicited proposals received and sponsored come from universities. Of the 75 solicited proposals from universities, 20 have been selected for CASIS sponsorship;[9] of the 43 solicited proposals from firms, 6 have been selected; and of the 5 solicited proposals from other government agencies, none have been selected.

From FY 2012 through FY 2014, 27 unsolicited projects were received from universities, 66 from firms, and 2 from other government agencies; 12 of the unsolicited university proposals, 40 of the unsolicited firm proposals, and both of the unsolicited proposals from other government agencies were selected for CASIS sponsorship.

The evaluation criteria for solicited and unsolicited proposals are similar, and according to CASIS they have and will continue to evolve over time. Basically, the evaluation process involves five stages:[10]

- **Operations Evaluation.** Expedited review by the CASIS Operations team to determine technical feasibility of the proposed project and achievability of the estimated budget and timeline.

- **Scientific Evaluation.** Evaluation by subject matter experts to score scientific merit and potential impact.

- **Economic Evaluation.** A two-pronged economic evaluation led by the CASIS economic staff, sometimes with external subject matter experts, to score potential tangible and intangible value.

- **Risk and Compliance Review.** Review by the CASIS Compliance team for regulatory and legal risks.

- **Final Determination.** The Executive Director, with input from the review process as well as advice from senior CASIS management, will perform the final prioritization and award determination.

---

8 CASIS's fiscal year (FY) ends September 30.

9 CASIS sponsorship includes a mix of direct and indirect costs as well as costs related to all of the operational, safety, and science verification tests that take place to ensure safety and maximize chances of successful function while on the ISS. According to CASIS, the approximate marginal cost for a firm experiment on the ISS is $1.54 million of which the firm contributes approximately between 15% and 30%. This cost estimate is based on a payload mass of 20 kgs, 18 months of payload development time, and 2 hours of crew time intervention, and the use of flight hardware. CASIS also estimates that NASA funding for programmatic support—overhead, infrastructure, and expertise that has built up over time that was required to initiate and maintain the ISS R&D program—is $7.4 million.

10 These criteria are described in each Request for Proposals. See also, *http://www.iss-casis.org/Opportunities/UnsolicitedProposals.aspx*.

TABLE 3.1: Solicited Proposals Received and Selected

SOLICITATION YEAR	RESEARCH AREA/DESCRIPTION	UNIVERSITY PROPOSALS RECEIVED/SELECTED	FIRM PROPOSALS RECEIVED/SELECTED	OTHER GOV'T AGENCY PROPOSALS RECEIVED/SELECTED
2012	**Material Science:** To exploit the space environment for testing of materials and devices. It is expected that applications will utilize the NanoRacks External Platform for development or testing of advanced sensors, electronics or materials that have commercial applications on Earth.[a]	5/1	5/1	0/0
2012	**Crystallography:** To exploit the microgravity environment to improve quality and yield of protein crystals and to establish the benefits of space science for understanding and applying crystallization methods.[b]	12/5	4/1	0/0
2013	**Stem Cells:** To investigate the impact of the spaceflight environment on the fundamental properties of mammalian stem cells.[c]	35/7	13/0	1/0
2014	**Remote Sensing:** To support use of facilities currently aboard the ISS to promote technology development and short-duration instrument demonstration (maximum of 90 days) for the specific purpose of remote sensing.[d]	12/5*	10/0	4/0
2014	**Enabling Technology to Support Science:** To solicit flight projects for enabling technologies for the development, testing, and/or utilization of new technologies, components and/or systems that will enable science-based investigations for Earth-based applications.[e]	5/1	7/2	0/0
2014	**Material Science:** To support flight research projects in the field of materials science for development and testing of materials and components that will lead to Earth-based applications.[f]	6/1	4/2	0/0
2014	**Energy Technology:** To use the ISS for studies of Earth for identifying or improving upon Earth-based energy applications such as energy capture, storage, or sustainability.[g]	underway	—	—

Information in this table came from the sources listed below and from personal correspondence with CASIS management. The statistics in the table are for solicited proposals. Step-1 proposals that were not invited or did not submit full step-2 proposals are not included. CASIS administers a two step-solicitation process that includes an abbreviated proposal first, which is then reviewed and either invited to submit a full step-2 proposal or not.

* Two of the five proposals are collaborating on a project but they are counted as two selections here as they were two separate proposals. Upon award, CASIS urged them to collaborate, and they have altered their plans to do so.

Sources:
a http://www.iss-casis.org/Portals/0/docs/RFP_MaterialsScience.pdf.
b http://www.iss-casis.org/Portals/0/docs/CASIS%20Request%20for%20Proposals-Crystallography%2006%2026%202012.pdf.
c http://www.iss-casis.org/Portals/0/docs/CASIS_RFP_Stem_Cell_v1.01.pdf.
d http://www.iss-casis.org/Portals/0/docs/CASIS_RFP_2013-3_Remote_Sensing_v.1.04.pdf.
e http://www.iss-casis.org/files/CASIS_2014-2_Enabling_Technology_022514_FINAL.pdf.
f http://www.iss-casis.org/files/CASIS_2014-4_RFP_Materials_Science.pdf.
g http://www.iss-casis.org/files/RFP_CASIS_2015-2_Earth_Observation_to_Benefit_Energy_Technology_v1-02.pdf.

Section 4. Survey-Based Information

Our model in Section 2 suggests that providing information about past R&D projects and experiments on the ISS might induce firms to increase their ISS R&D. We wanted to collect data to test this idea. Specifically, we wanted to see what information firms used when proposing to conduct ISS R&D. And we wanted to see whether there was information firms wished they had had when making proposals.

We developed three survey instruments and requested, through NASA, that CASIS solicit responses to the survey questions. One instrument was administered to firms selected in FY 2014 and FY 2015 to conduct ISS R&D. A second instrument was directed to firms that submitted a proposal in FY 2014 or FY 2015 but were not selected. The third instrument was given to firms that had already completed ISS R&D. The three survey instruments are reproduced in Tables 3.2, 3.3, and 3.4, respectively.

We did not seek Office of Management and Budget (OMB) clearance under the Paperwork Reduction Act to administer these surveys because of time constraints imposed by NASA on this project. Thus, we asked CASIS to randomly select a maximum of 9 firms to survey from each of the three categories. CASIS administered the appropriate survey to 9 of the 52 firms selected in FY 2014 or FY 2015 to conduct ISS R&D projects and received responses from all of them;[11] they also administered the appropriate survey to 9 of the 139 firms that were not selected and received 6 responses.[12] To date, 7 firms have completed R&D projects on the ISS; CASIS received survey responses from all 7. We discuss the responses from each survey below.

4.1 Responses from Firms Selected to Conduct ISS R&D in FY 2014 and FY 2015

Firm-identifying information on the survey instruments was generally masked by CASIS. However, a number of the open-ended responses contain descriptors of the surveyed firms' projects. For the sake of confidentiality, we do not report all these responses fully. Rather, we summarize them and selectively include non-identifying comments in our summary interpretations.

---

11 Of the 52 firms that were selected to conduct ISS R&D in FY 2014 or FY2015 year-to-date, 18 had submitted solicited proposals. Of the 9 responses, 0 were from firms that submitted solicited proposals.

12 Of the 139 firms that were not selected to conduct ISS R&D in FY 2014 or FY 2015 year-to-date, 110 had submitted solicited proposals. Of the 6 responses, 5 were from firms that submitted solicited proposals. Of the 5 solicited proposals, 4 failed the Scientific Evaluation stage of the review process. The 5th proposal wanted to modify a piece of ISS hardware that NASA did not want to modify. The non-solicited firm failed the Operations Evaluation stage of the review process.

TABLE 3.2: Survey Instrument for Commercial Firms that Were Selected in CASIS FY 2014 or FY 2015 to Conduct ISS R&D Projects

---

Thank you for participating in this survey for our NASA project on commercializing the ISS. Please be assured that your responses will remain confidential and only a summary of our findings will be reported to NASA in our final report without attribution to you or your firm.

1. We are interested in the sources of information that influenced your decision to propose an R&D project to be conducted on the ISS. Briefly, please list the 3 most influential sources from most to least influential.

    a.

    b.

    c.

2. We are also interested in additional types of information you would have liked to have had when deciding to propose an R&D project to be conducted on the ISS. Briefly, please list what that information might have been from most to least important.

    a.

    b.

    c.

Please respond to the following statements using a 7-point Likert scale of 1 = strongly disagree, 2 = moderately disagree, 3 = disagree, 4 = neither disagree nor agree, 5 = agree, 6 = moderately agree, and 7 = strongly agree.

3. My firm expects to be able to commercialize the technology that is expected to result from the ISS R&D. _____

    Mean = 5.08   Median = 5.50   Range = 2–7

(We use the term *technology* to refer to the application of new knowledge, learned through science, to some practical problem(s). R&D is critical to this application process. In contrast, we think of an innovation as technology put into use or commercialized.)

4. My firm expects to conduct additional R&D projects on the ISS. _____

    Mean = 4.42   Median = 4.00   Range = 3–7

5. My firm would have proposed an ISS R&D project sooner if it had had information about other firms' successes on the ISS. _____

    If such information would have made a difference, please indicate what form it could have taken (e.g., information about other firms' returns).

    Mean = 3.67   Median = 4.00   Range = 1–7

Question 1 (see Table 3.2) asked firms to state what information influenced their decision to propose a project. In response, most selected firms (6 of 9) mentioned the importance of prior formal and informal interactions with CASIS and NASA. Several respondents (3 of 9) also cited the existing scientific literature as significant in their decision process. Indeed, one firm noted that the literature related to its experiment was the most influential source of information. A second firm responded that it had "performed a general review of the literature around the experiments [it was] looking at" and that this review was the most important form of information. A third respondent noted the particular influence of "technical knowledge gained" (presumably gained as background for the application process).

Question 2 asked firms what additional types of information would have been useful to them when deciding to propose an ISS R&D project. In their answers, a large majority of firms (7 of 9) mentioned wanting more data about previous R&D experiments and about the ISS environment. Here are some representative responses:

- More information on past, present and planned space experiments, particularly around the area of [our research].

- More information on what is feasible for experiments on the ISS.

- A better way to search for experiments that have been done...and what they set out to explore and what they learned.

- Planning information of other previous firm projects, to better understand how long is the time frame between the first contact with CASIS to the project approval and completion.

- Typical final costs and schedules of previously successful projects most similar to the one I was proposing.

- More background materials on experiment design and implementation.

- It would have been nice to understand the different hosting options and the costs and details of those options for ISS flight experiments.

We think the fact that a large proportion of selected firms surveyed wished they had had more information about previous R&D experiments is the strongest empirical confirmation of our theoretical suggestion that such information would likely expand R&D (albeit our sample size of 9 was small). It should be noted, however, that we uncovered little evidence that this information would encourage R&D to be done earlier than otherwise. Specifically, we also asked (Question 5 in **Table 3.2**) firms to respond to the following statement using a 7-point Likert scale of 1 = strongly disagree to 7 = strongly agree: *My firm would have proposed an ISS R&D project sooner if it had had information about other firms' successes on the ISS.* The mean response was 3.67 and the median response was 4.00.[13] That is, the response was, on average, neutral.

We also got a fairly neutral responses to other questions: Question 3 asked firms to respond to the statement: *My firm expects to be able to commercialize the technology that is expected to result from the ISS R&D.* The average response was 5.08 and the median

---

13 There were 9 firms that responded to this survey, but one firm will be conducting 4 experiments. Thus, there were 12 numerical responses in all.

response was 5.50. That is, firms, on balance, seem to have fairly modest expectations about being able to commercialize their ISS R&D. Question 4 (in Table 3.2) asked firms to respond to the statement: *My firm expects to conduct additional R&D projects on the ISS.* The average response was 4.42 and the median response was 4.00.[14] In other words, firms are apparently quite uncertain about the prospect doing further ISS R&D.

TABLE 3.3: Survey Instrument for Commercial Firms that Were Not Selected in CASIS FY 2014 or FY 2015 to Conduct ISS R&D Projects

---

Thank you for participating in this survey for our NASA project on commercializing the ISS. Please be assured that your responses will remain confidential and only a summary of our findings will be reported to NASA in our final report without attribution to you or your firm.

1. We are interested in the sources of information that influenced your decision to propose an R&D project to be conducted on the ISS. Briefly, please list the 3 most influential sources from most influential to least influential.

    a.

    b.

    c.

2. We are also interested in additional types of information you would have liked to have had when deciding to propose an R&D project to be conducted on the ISS. Briefly, please list what that information might have been from most important to least important.

    a.

    b.

    c.

3. Do you expect to submit another proposal to conduct R&D on the ISS? _____ (yes/no) If "no" what additional information would your firm need to submit another proposal?

    a.

    b.

    c.

---

[14] These two responses appear to be related. For example, Pearson's correlation coefficient between responses to Question 3 and responses to Question 4 is 0.52, and it is significant at the 0.10 level; and Spearman's rank order correlation coefficient is 0.63, and it is significant at the .05 level. Prior to beginning their ISS R&D, those firms that expect to commercialize the technology from their ISS R&D are also those firms that expect to conduct additional R&D projects on the ISS.

TABLE 3.4: Survey Instrument for Commercial Firms that Completed ISS R&D Projects

Thank you for participating in this survey for our NASA project on commercializing the ISS. Please be assured that your responses will remain confidential and only summary data will be reported to NASA in our final report without attribution to you or your firm.

1. Briefly, what information was/could have been helpful in your decision to undertake this ISS R&D project?

   a.

   b.

   c.

In the statements below we use the term technology to refer to the application of new knowledge, learned through science, to some practical problem(s). R&D is critical to this application process. In contrast, we think of an innovation as technology put into use or commercialized.

Please respond to the following statements using a 7-point Likert scale of 1 = strongly disagree, 2 = moderately disagree, 3 = disagree, 4 = neither disagree nor agree, 5 = agree, 6 = moderately agree, and 7 = strongly agree.

If this project alternatively could have been conducted as an on-Earth R&D project, please respond only to statements 2, 3, 4, and 5 below.

However, if this project, for technical reasons, could not possibly have been conducted as an on-Earth R&D project regardless of cost, please respond only to statements 6, 7, and 8.

2. The depth/breadth of research success realized from this ISS R&D project was greater than if the project had been conducted as an on-Earth R&D project. ____ $n = 4$

   Mean = 4.50   Median = 5.00   Range = 2–6

3. Once this project began, it took a longer time to complete than if it had been conducted as an on-Earth R&D project. ____ $n = 4$

   Mean = 6.50   Median = 6.50   Range = 6–7

4. Taking into account the research success of this project, and the time and cost it took to complete it, the total expected returns from a commercialized version of this technology will be greater than if the project was conducted as an on-Earth R&D project. ____ $n = 4$

   Mean = 4.00   Median = 4.00   Range = 1–7

5. The results of this research have resulted in a "go" decision for the next phase of R&D. ____ $n = 4$

   Mean = 3.25   Median = 3.00   Range = 1–6

Questions 6 through 8 ask for a comparison between this ISS R&D project and a previously completed on-Earth project that concerned a similar technology and was of comparable scale and scope.

6. The depth/breadth of research success realized from this ISS R&D project was greater than from a previously completed on-Earth project that concerned a similar technology and was of comparable scale and scope. ____ $n = 6$

   Mean = 4.83   Median = 4.50   Range = 3–7

> 7. Once this ISS R&D project began, it took a longer time to complete than a previously completed on-Earth project that concerned a similar technology and was of comparable scale and scope. _____ n = 5
>
> Mean = 6.60   Median = 7.00   Range = 6–7
>
> 8. Taking into account the research success of this ISS R&D project, and the time and cost it took to complete it, the total expected returns from a commercialized version of this technology will be greater than those from a previously completed on-Earth project that concerned a similar technology and was of comparable scale and scope. _____ n = 6
>
> Mean = 4.83   Median = 4.00   Range = 3–7

4.2 Responses from Firms Not Selected to Conduct ISS R&D in FY 2014 and FY 2015

Like most selected firms, some non-selected firms mentioned prior interactions with CASIS and NASA as important to their decision to propose a project; however, only half of them (3 of 6 did so), by contrast with two thirds (6 of 9) of the selected firms. None of the non-selected firms cited the scientific literature as crucial to their decision to make a proposal, whereas 33 percent of the selected firms did.

As for additional types of information that they wished they had had (Question 2 in **Table 3.3**), none of the non-selected firms mentioned that they would have benefitted from more information about past R&D projects and experiments. Some firms (3 of 6) said that they would have liked more information about to the application process. Such information included:

- technical details regarding integration support in the call for proposals;
- clearer picture of selection criteria and
- better [understanding of] the recommended sequence of events from identification of proposal opportunity, to proposal selection through to sample recovery at the mission's conclusions.

Also, several firms (3 of 6) stated that they would have liked to know more about additional funding. These firms responded that the following information would have been useful:

- information about other possible sources of funding to supplement CASIS award;
- alternative funding sources [if] not selected; and
- what sources [are] available to provide [for a] shortfall between funding provided and funding required to complete the mission.

4.3 Comparison of Selected and Non-Selected Firm Responses

Table 3.5 summarizes the difference between selected and non-selected firms concerning information that influenced their decision to apply to NASA. Both groups mentioned interactions with NASA and CASIS, but only the selected firms emphasized the previous scientific literature.

There was an even sharper difference between the two groups over "information they wished they had had" (see Table 3.6). The selected firms stressed past R&D experiments and the ISS environment; the non-selected firms cited the application process and additional funding opportunities.

We can only speculate on the reasons for these differences, but perhaps non-selected firms' responses were, in part, a way of rationalizing their failure to be selected.

TABLE 3.5: Comparative Findings About Information that Influenced Firms' Decision to Propose an ISS R&D Project

INFORMATION	SELECTED FIRMS	NOT SELECTED FIRMS
Formal/informal interactions with CASIS and/or NASA	✓	✓
Academic/scientific literature related to ISS R&D project	✓	

TABLE 3.6: Comparative Findings About Information that Firms Would Like to Have Had When Deciding to Propose an ISS R&D Project

INFORMATION	SELECTED FIRMS	NOT SELECTED FIRMS
About past R&D experiments	✓	
About ISS environment	✓	
About application process		✓
About additional funding opportunities		✓

4.4 Responses from Firms that Completed ISS R&D Projects

To date, 7 firms have completed ISS R&D projects. Each firm was asked (Question 1 in Table 3.4): *What information was/could have been helpful in your decision to undertake this ISS R&D project?*

Below is a representative sample of responses. In our opinion, all of these touch on information that might be relevant to a firm's decision to expand its ISS R&D:

- positive results from previous microgravity [experiments like ours];
- better connection with NASA and ISS administrators around project management;
- knowing that there was a system in place to perform our specific type of experiment;

- helpful to know in advance which other companies that we would need to work with to make the experiments a reality;
- support material that show the ISS as a research resource at a level deeper than what is currently presented. Working through safety and integration processes seems that you need to have insider knowledge for a lot of processes;
- estimated costs; and
- knowledge of flight hardware and reasonable expectations of success.

Only one firm (1 of 7) mentioned that information about past R&D projects and experiments would have been beneficial when deciding to undertake its ISS R&D project: "Positive results from previous [experiments like ours]" would have been helpful. Of course, this is not terribly surprising, because at the time that these firms proposed their projects, there were few prior experiments from which they could have learned. They were, in a real sense, the pioneers of ISS R&D.

The remaining survey questions were intended to find out what types of information from R&D projects might be useful to subsequent project proposers. These questions were divided into two categories.

Questions in the first category were prefaced by: *If this project alternatively could have been conducted as an on-Earth R&D project.* Four of the 7 firms responded to these questions; their answers were given as numbers along the 7-point Likert scale and are summarized in **Table 3.4**.

The results show that there is general agreement that the ISS R&D project took *longer to complete than if it had been conducted as an on-Earth project* (Question 3 in **Table 3.4**). The mean and median responses were 6.50.

There was some agreement that *the depth/breadth of research success realized from the ISS R&D project was greater than if the project had been conducted as an on-Earth project* (Question 2 in **Table 3.4**). The mean response was 4.50 and the median response was 5.00.

We got a neutral response to the following question (Question 4 in **Table 3.4**): *Taking into account the research success of this project, and the time and cost it took to complete it, the total expected returns from a commercialized version of this technology will be greater than if the project was conducted as an on-Earth R&D project.* The mean and median responses were 4.00.

Finally, firms slightly disagreed with the idea that their results from the completed R&D project were sufficient for *a "go" decision for the next phase of R&D* (Question 5 in **Table 3.4**). The mean response to Question 5 was 3.25 and the median response was 3.00.

The second category of questions asked each firm *for a comparison between its ISS R&D project and a previously completed on-Earth project that concerned a similar technology and was of comparable scale and scope.*

There was strong agreement among the 5 respondents that the ISS R&D project took a *longer time to complete* (Question 7 in **Table 3.4**). The mean response was 6.60 and the median response was 7.00.

There was mild agreement among the 6 respondents that the *research success* from the ISS R&D project was greater than from a comparable on-Earth project (Question 6 in **Table 3.4**). The mean response to Question 6 was 4.83 and the median response was 4.50.

Finally, there was also modest agreement among the 6 respondents that the *total expected returns from a commercialized version of this technology* will be greater than those from a comparable on-Earth project (Question 8 in **Table 3.4**). The mean response was 4.83 and the median response was 4.00.

All in all, apart from the fact that ISS R&D takes longer than Earth-bound R&D, we do not feel we learned a great deal from these firms about what information might be useful for their successors to have.

Section 5. Recommendations to NASA

Our recommendation to NASA is to provide candidate firms with more information about past R&D projects and experiments conducted on the ISS. This recommendation has an economic logic that we developed in **Section 2**. It is also backed up by survey data, in particular, by the responses from firms recently selected to conduct ISS R&D.

To implement our recommendation we suggest that NASA, through CASIS, systematically collect information about past R&D projects and experiments conducted on the ISS and make that information available to all R&D firms. Our own survey of firms with completed projects did not clearly identify what the nature of that information should be. But we are not experts on R&D in LEO. Thus, it might be a good idea to assemble an advisory group to formulate survey questions for firms during and after their ISS projects are completed. Over time, a database of information about experiences from ISS R&D could be constructed, perhaps segmented by technology.[15] Firms contemplating ISS R&D could access this database before deciding how much to invest. We believe that this information will increase R&D on the International Space Station and thereby help expand the commercial possibilities of low Earth orbit.

---

15 The Advanced Technology Program (ATP) that operated at the National Institute of Standards and Technology (NIST) from 1990 to 2007 funded firms both individually and as part of a research joint venture to conduct R&D to develop pre-competitive generic technologies. The General Terms and Conditions for these awards has Technical Progress Reporting requirements and Business and Economic Reporting requirements. See *http://www.atp.nist.gov/atp/2007_atp_gen_terms_cond_9_24_07_final.pdf*. The Business and Economic Reporting requirements included an annual survey, end-of-project survey, and often a post-project survey. ATP made this information publicly available. Thus, not only might this information have influenced the decision of candidate firms to apply for ATP support but it also created a database for studying economic aspects of ATP's activities. See Powell (1998) for a discussion about the development and use of such a database.

References

NASA (2007). "NASA Report to Congress Regarding a Plan for the International Space Station National Laboratory," *http://www.nasa.gov/pdf/181149main_ISS_National_Lab_Final_Report_rev2.pdf,* accessed on February 1, 2015.

NASA (2011). "Cooperative Agreement Notice: ISS National Laboratory Management Entity," Reference Number: NNH11SOMD002C, February 14.

Powell, Jeanne (1998). "Pathways to National Economic Benefits from ATP-Funded Technologies," *Journal of Technology Transfer* 23(2): 21–32.

Smith, Marcia S. (2001). "NASA's Space Station Program: Evolution and Current Status," Testimony before the House Science Committee, April 4.

The White House (2011). "Presidential Memorandum: Accelerating Technology Transfer and Commercialization of Federal Research in Support of High-Growth Businesses," October 28.

CHAPTER 4

Venture Capital Activity in the Low-Earth Orbit Sector

Josh Lerner[1,2]
Ann Leamon[2]
Andrew Speen[2]

Executive Summary

NASA seeks to understand the challenges and opportunities inherent in the economic development of the low-Earth orbit (LEO) sector.[3] This sector comprises private companies seeking to commercialize LEO-related technologies. The Center for the Advancement of Science in Space (CASIS) has been managing the International Space Station (ISS) U.S. National Laboratory since 2011 and supports space-related projects intended for terrestrial benefit.

We assess the LEO sector with respect to ISS commercialization activities as well as LEO activity outside of direct NASA-supported efforts. More specifically, we explore these trends in LEO and their potential for innovation through the lens of venture capital (VC). Because venture capitalists invest in innovative, high-tech companies, the degree of viable LEO activity in general and ISS commercialization specifically would likely be reflected in trends of VC-backed companies addressing opportunities in LEO. We track LEO VC activity in three ways:

- First, we qualitatively analyze VC interest in the LEO sector and the degree to which it has waxed and waned over time. To do so, we first review academic studies, market forecasts, conference reports/proceedings/lectures, NASA reports,

---

[1] Harvard Business School.
[2] Bella Research Group.
[3] LEO is technically defined as an orbit at an altitude between 80 km and 2000 km above the Earth's surface. See, National Aeronautics and Space Administration. *Ancillary Description Writer's Guide: Global Change Master Directory*, 2016.

Disclaimer: The views and opinions of the authors do not necessarily state or reflect those of the U.S. Government or NASA.

press releases, news articles, and VC and corporate blogs. We also conduct two case study investigations of LEO companies—one of which has utilized the ISS, the other has not, to date—to highlight emerging trends in the sector. To support these analyses, we interview a small group of venture capitalists and an aerospace lawyer active in LEO to better understand their perspectives on the risks and opportunities of LEO at its current state.

We find that the LEO sector has historically been unattractive to venture capitalists due to the prohibitively infrequent launch and expensive opportunities; high capital requirements for satellite designs; regulatory barriers, and a lack of commercially viable applications. Over the last 5 to 10 years, however, VC interest has emerged as a result of technological developments enabling medium- and high-resolution images from miniature satellites, more frequent miniature satellite launch opportunities, and rising interest of "big data" analytics for various types of monitoring activities (e.g., agricultural, maritime, mining) and financial trading intelligence. Additionally, "emerging"—although largely unproven—advances in higher frequency, miniature satellite-dedicated launches reflect a more conducive environment for LEO startups to access space.

- Second, we empirically examine trends in U.S. VC activity (i.e., equity invested and deal counts) related to LEO companies. Our analysis confirms an emergence of VC activity and suggests that the focus is on launch vehicles and miniature earth observation/remote sensing satellites. In addition, we find that the primary VC backers of LEO companies seem to have a history of investing in Internet- and software-related companies, which suggests that LEO-related firms may exhibit features similar to these more traditional VC subsectors.

- Third, we analyze the performance of VC-backed LEO firms. We first look to see if any successful exits (i.e., initial public offering (IPO), acquisition) were achieved, as well as if any bankruptcies have been reported in the industry. Because exit opportunities have been limited due to the relative youth of LEO-related startups, we also look to see if VC-backed firms achieve multiple rounds of financing. We find that over half of the companies that received an initial VC investment between 2008 and 2014 have received additional rounds of financing as of January 2015. Importantly, this subset of LEO companies has also seen one high-profile exit—Google's acquisition of SkyBox for $500 million—and, a major valuation success with SpaceX, a private company that was valued at over $10 billion when it raised money in January 2015.[4] No bankruptcies have been identified.

Our findings collectively suggest that venture capitalists are increasingly interested in LEO-related companies. The majority of this interest stems from launch and miniature satellite startups. We note that while some miniature satellite companies

---

4 SpaceX raised $1 billion in a round led by Google and Fidelity for roughly 10 percent equity.

deploy via ISS, others avoid the ISS. Companies servicing miniature satellites, such as dedicated launch vehicles and propulsion systems, also appear to be of interest, as evidenced by recent deals from notable investors in each subsector.

Outside of the launch and miniature satellite domains, however, LEO entrepreneurship that directly relates to technology developed or researched on the ISS—e.g., microgravity enabled products (materials, pharmaceuticals, vaccines)—appears to have attracted little interest from venture capitalists to date. This finding reflects most directly a scarcity of startup companies formed to exploit microgravity-related technology. We hypothesize that entrepreneurship has been restrained by the logistics of LEO manufacturing—most notably, prohibitive launch costs and infrequent, sometimes unpredictable launch opportunities—in spite of clear developments in both of these areas. Another barrier seems to be a general lack of awareness on the part of venture investors that the ISS goes beyond the realm of basic science and can serve as a platform for entrepreneurial enterprises.

In light of this analysis we develop four recommendations on how NASA and CASIS can promote VC involvement in the sector and spur economic development via the ISS.

1. Expand efforts to raise awareness among LEO-oriented entrepreneurs and relevant angel networks/VC groups of the ISS, its benefits, and how to utilize it for applied commercial research. NASA and CASIS should focus these efforts on disseminating information regarding (i) the intersections of the ISS with such industries as biotechnology and miniature satellites, (ii) the steps needed to use the ISS efficiently, and (iii) the relevant successes to date.

2. Set up a committee of venture capitalists to advise NASA and CASIS management on strategy of private sector ISS involvement. Venture capitalists could offer a fresh strategic perspective on ISS commercialization by (i) identifying current trends in LEO entrepreneurship, (ii) providing subtle knowledge of the challenges entrepreneurs encounter with the ISS, and (iii) further raising VC awareness of the ISS's commercial applications.

3. Continue to collaborate with relevant angel networks. Given academic research suggesting complementarities between VC and angel investments, we suggest continued relationships with specialist angel networks in the fields applicable to LEO entrepreneurship (e.g., biotechnology, materials, data and analytics).

4. Continue to partner with accelerator programs and offer additional funds to ease the "proof-of-concept" process for accelerated startups. Accelerator programs can help mature the "formation-stage" LEO companies that may not be appealing to traditional venture capitalists or even some angel investors. In fact, graduates of accelerator programs have been found to attract VC investments. With CASIS supporting ISS-oriented accelerated startups with timely and inexpensive access to the ISS, such companies will more likely garner VC attention.

Section 1. Introduction

In this analysis, we examine the private LEO sector through the lens of venture capitalists. We ask the following questions:

- What have traditionally been and what are currently the main opportunities and challenges in the LEO sector for venture capitalists?
- What is the extent of VC activity in LEO—especially with respect to LEO companies utilizing the ISS—and how has it changed over time?
- In what specific areas of LEO do venture capitalists invest and how do these areas intersect with the ISS?
- What has been the performance to date of VC-backed LEO firms?

We investigate the evolution of the sector since the late 1980s and identify entrepreneurial trends shaping the sector today. We begin, however, with a brief overview of why we explore the private LEO sector and, in particular, why we do so from the perspective of venture capitalists.

1.1 Why Examine the Private LEO Sector?

Commercial activities on the International Space Station (ISS) are fundamental to its very existence. The 1984 amendment to the National Aeronautics and Space Act of 1958—the law that created NASA—declared that commercialization was a necessary ingredient (an explicit statutory requirement) to promote the general welfare and security of the United States via aeronautical and space activities.[5] This sentiment was further encouraged by subsequent pieces of legislation, such as in the Commercial Space Launch Amendments Act of 2004 and NASA Authorization Acts.[6]

Given the statutory emphasis on private sector participation in LEO and on the ISS, this chapter explores commercial activity in LEO and the extent to which the ISS is viewed as a viable source of start-up opportunities. In particular, we look at

---

[5] National Aeronautics and Space Administration Authorization Act, 1985, Pub. L. No. 98-361, Sec. 110(a), 98 Stat. 426 (July 16, 1984). The law stated: "Congress declares that the general welfare of the United States requires that the National Aeronautics and Space Administration... seek and encourage, to the maximum extent possible, the fullest commercial use of space." See also, "Commercialization of Space: Commercial Space Launch Amendments Act of 2004." *Harvard Journal of Law & Technology* 17, no. 2 (Spring 2004): 622.

[6] See, Commercial Space Launch Amendments Act of 2004, Pub. L. No. 108-492, 108th Congress, 118 Stat. 3974 (Dec. 23, 2004). Examples of NASA Authorization Acts include: National Aeronautics and Space Administration Authorization Act of 2005, Pub. L. No. 109-155, 109th Congress, 119 Stat. 2895 (Dec. 30, 2005); National Aeronautics and Space Administration Authorization Act of 2010, Pub. L. No. 111-267, 111th Congress, 124 Stat. 2805, Sec. 202(b)(1) (Oct. 11, 2010); and National Aeronautics and Space Administration Authorization Act of 2014, H.R. 4412, 113th Congress, Sec. 211(a)(2) (June 23, 2014).

VC activity in the LEO space. Broadly speaking, VC activity in the LEO sector would indicate the existence of innovative, high-tech companies—not only technologies—in the sector. Before exploring VC activity in LEO, though, we provide an overview of the venture capital model of financing to better understand our rationale for this study.

1.2 Why Examine the Private LEO Sector Via Venture Capital Activity?

Venture capital is a key financing source for high-quality entrepreneurial firms. To be sure, VC is a relatively small financial institution. In each of the five years from 2009 to 2013, the National Venture Capital Association (NVCA) Yearbook reported that fewer than 1,500 companies received VC for the first time in the United States.[7] This is a very small fraction—roughly 1 in 400, or 0.25%—of the roughly 600,000 firms (with employees) that are started each year.[8] But VC funding has had a disproportionate impact on the U.S. economy. Samuel Kortum and Josh Lerner found that VC was roughly three times more powerful than corporate R&D in stimulating patenting.[9] Furthermore, Steve Kaplan and Josh Lerner found that roughly 60 percent of the "entrepreneurial" IPOs from 1999 to July 2009 were venture-backed, despite the small fraction of all firms that receive venture funding at all.[10]

By way of background, the VC model is built around financing innovative ventures with uncertain futures. As noted by Paul Gompers and Josh Lerner, "[v]enture investors typically concentrate in industries with a great deal of uncertainty, where the information gaps among entrepreneurs and investors are commonplace."[11] William Kerr, Ramana Nanda, and Matthew Rhodes-Kropf in fact describe VC investors

---

[7] Thomson Reuters. *National Venture Capital Association Yearbook 2014*. Arlington: NVCA, 2014, p. 55.

[8] This figure is calculated as average "employer births" from 2005 to 2010. See U.S. Small Business Administration, 2012, *The Small Business Economy*, Washington, DC: SBA.

[9] Samuel Kortum and Josh Lerner. "Assessing the Contribution of Venture Capital to Innovation." *Rand Journal of Economics* 31, (2000): 674–692, 675. A positive contribution of venture capitalists to their portfolio companies is also found in Shai Bernstein, Xavier Giroud, and Richard Townsend. "The Impact of Venture Capital Monitoring: Evidence from a Natural Experiment." (February 23, 2014).

[10] The authors also noted that in only two of these 11 years have fewer than 50 percent of IPOs been VC-backed. "Entrepreneurial IPOs" defined as non-financial, non-reverse LBO, non-REIT, non-SPAC IPOs. See Steven N. Kaplan and Josh Lerner. "It Ain't Broke: The Past, Present, and Future of Venture Capital." *Journal of Applied Corporate Finance* 22, no. 2 (2010): 36–47.

[11] Paul Gompers and Josh Lerner. The Venture Capital Cycle. 2nd ed. Cambridge, MA: MIT Press, 2004. See also, Paul A. Gompers. "Optimal Investment, Monitoring, and the Staging of Venture Capital." *The Journal of Finance* 50, no. 5 (December 1995): 1461–1489.

as "conducting a portfolio of tests across...highly uncertain ideas."[12] VC firms pool funds from investors and offer an illiquid, long-term investment in a portfolio of these young, unproven companies. The investors typically include insurance companies, foundations, endowments, and public and private pension funds and act as the limited partners (LPs) in a fund. The equity stake that VC firms receive in exchange for capital in portfolio companies offers potentially lucrative rewards that counterbalance the high investment risks. At the exit, the LPs receive their invested capital and the profits are split; the LPs typically receive 80 percent and the VC firm (the general partner, or "GP") 20 percent. The LPs also pay fees to the GPs to cover the costs of doing business.

To mitigate risk, venture capitalists perform extensive due diligence to identify companies best suited for successful exits. Due diligence investigation typically involves a look into management team capabilities, technological innovation, and market potential, as well as various other characteristics of the company.[13] Research suggests that for every 100 business plans submitted to VC groups only one receives funding on average.[14]

To further mitigate risk and align interests, venture capitalists stage financing by funding companies based on the milestones they achieve. This approach acts as a control mechanism to keep entrepreneurs tightly focused on creating value for the firm. It also allows the venture capitalists to evaluate the firm's progress, especially in light of possibilities that entrepreneurs may pursue projects with high private returns but low monetary returns for investors.[15] In addition, by allowing investors to terminate funding in light of negative signals (e.g., failed technologies, unfavorable patent decisions, and so on), venture capitalists can pursue projects that would be too risky in an "all-or-nothing bet."[16] Sequential investments further allow venture investors to adjust investment paths as information emerges that enables more accurate estimates of the venture's probability of success.[17]

---

12 William R. Kerr, Ramana Nanda, and Matthew Rhodes-Kropf. "Entrepreneurship as Experimentation." *The Journal of Economic Perspectives* 28, no. 3 (2014): 29.

13 Andrew Metrick and Ayako Yasuda. *Venture Capital & the Finance of Innovation*. 2nd ed. Wiley, 2011: 139–141.

14 Thomson Reuters. *National Venture Capital Association Yearbook 2014*. Arlington: NVCA, 2014, 7.

15 Paul A. Gompers. "Optimal Investment, Monitoring, and the Staging of Venture Capital." *The Journal of Finance* 50, no. 5 (Dec., 1995): 1461–1489.

16 This theory is described in William R. Kerr, Ramana Nanda, and Matthew Rhodes-Kropf. "Entrepreneurship as Experimentation." *The Journal of Economic Perspectives* 28, no. 3 (2014): 25–48, 28.

17 This idea is discussed in, Dirk Bergemann, Ulrich Hege, and Liang Peng. "Venture Capital and Sequential Investments." Cowles Foundation Discussion Paper no. 1682R (March 2009). The authors explain, "...learning about the expected final value of the failure probability is incorporated in all subsequent investment decisions [of venture capitalists]. If there is a positive news update then the value of the project increases as well as the investment flow" (p. 3).

It is important to emphasize that venture capitalists' foremost goal is to maximize financial return by exiting their portfolio companies via initial public offering (IPO) or acquisition. Given an average fund lifetime of roughly 10 years, venture capitalists consider exit options, as well as the expected time required to exit, prior to investment.[18] As a result, although venture capitalists do invest in "raw startups"—companies composed of a few talented individuals and an idea—they generally avoid investing in raw technology without a management team at all. Most U.S.-based VC firms seek to invest in high-tech companies at the point of rapid growth potential.[19] According to the NVCA, only 11 percent of companies receiving VC financing in 2013 were considered "non-high technology."[20] Venture capitalists are particularly interested in software and biotechnology, which respectively represented 37 percent and 15 percent of investments in 2013.[21]

With this background of VC financing, **Section 2** continues with a qualitative overview of LEO-related entrepreneurship to examine the extent of VC interest in the sector in general and in ISS-related companies specifically. In **Section 3**, we empirically track VC activity in the U.S. LEO sector and describe specific characteristics of such deals (e.g., frequency of different subsectors and primary investors involved). In **Section 4** we look at the extent of subsequent funding rounds and exits in our sample of VC-backed LEO firms. **Section 5** offers a set of recommendations on how to spur VC interest in the sector. **Section 6** concludes with some final thoughts.

Section 2. Qualitative Overview of VC Involvement in LEO

In this section we review the literature related to VC investment in the LEO sector. Because there is limited academic material related specifically to VC interest in LEO, we also examine market forecasts, conference reports/proceedings/lectures, news reports, and VC and corporate blogs. This overview includes two case studies, which distill key themes relating to the current VC perspective on the LEO sector, as well as insights from interviews with venture capitalists and an attorney involved in the sector. We note that our review is not exhaustive in nature and instead aims to capture the central challenges and opportunities in the sector at large and for the ISS specifically from the perspective of venture capitalists.

---

18 For a discussion of exit dynamics for U.S. VC funds, see Pierre Giot and Armin Schwienbacher. "IPOs, Trade Sales and Liquidations: Modelling Venture Capital Exits using Survival Analysis." *Journal of Banking and Finance* 31, no. 3 (2007): 679–702.

19 Andrew Metrick and Ayako Yasuda. *Venture Capital & The Finance of Innovation.* 2nd ed. Wiley, 2011, p. 6.

20 Thomson Reuters. *National Venture Capital Association Yearbook 2014.* Arlington: NVCA, 2014.

21 Thomson Reuters. *National Venture Capital Association Yearbook 2014.* Arlington: NVCA, 2014.

2.1. Terminology

Before proceeding, it is important to clarify the terminology used throughout this chapter.

A. Orbit Altitudes

At the most general level, orbit altitudes can be divided into four types. Low Earth orbit (LEO) is between 80 km and 2,000 km; medium Earth orbit (MEO), also known as intermediate circular orbit, is between 2,000 km and 35,786 km, geosynchronous orbit (GSO) is at 35,786 km; and high Earth orbit (HEO) is above 35,786 km.[22] We note that the ISS is situated in LEO at an orbit altitude of 370–460 km. In this chapter, we focus on the commercial activity in LEO and on the ISS specifically.

B. Miniature Satellite Classes

Satellites are generally classified according to their mass and/or structure. We use "miniature satellites" as an umbrella term encompassing nanosatellites (1–10 kg), microsatellites (10–100 kg), or small satellite (100–500 kg). While miniature satellites technically also include femtosatellites (10–100g) and picosatellites (less than 1 kg), these classes are of minimal relevance to the chapter. In addition, "CubeSats" typically refer to nanosatellites built by 10×10×10-centimeter blocks, or units, each weighting between 1 kg and 1.33 kg. CubeSats are typically configured as one (1U), two (2U), or three (3U) units in length. CubeSats adhere to standard launch containers mounted to launch vehicles, which promotes flexibility in launch options for developers.[23]

C. Payload Classes

Similarly, commercial cargo, or the "payload" of a launch vehicle (satellites, research experiments, etc.), can be divided into two classes: primary and secondary. A primary payload fills the majority of vehicle capacity and determines the launch schedule and mission parameters, such as the orbit altitude. A secondary payload "hitches" a ride with the primary payload on a launch vehicle that has excess capacity, but the group that provides the secondary payload typically possesses no authority over mission dates or parameters.

---

22 For more information, see NASA. Ancillary Description Writer's Guide: Global Change Master Directory. 2015, *http://gcmd.nasa.gov/add/ancillaryguide/platforms/orbit.html*.

23 For a discussion of CubeSats, see, Michael Swartwout. "The First One Hundred CubeSats: A Statistical Look." *Journal of Small Satellites* 2, no. 2 (2013): 213–233.

2.2 VC Interest in LEO

We begin with an overview of the historic challenges faced by entrepreneurial startups and VC involvement in the LEO sector. We track the progression of these challenges in the following section with an investigation of the current (post-2010) VC perspective on entrepreneurial activity in LEO.

Historic Challenges of the LEO Sector for Venture Capitalists

Entrepreneurial interest in LEO has historically been constrained by a number of logistical, technical, and regulatory issues. Early studies that surveyed a number of actors in space-related industries between 1988 and 1999 found that high costs associated with launch and insurance, scarcity of managerial experience, limited market size, and long development times (relative to the life cycle of VC funds) collectively inhibited VC involvement in the sector.[24] Below, we more closely examine some of the key barriers to LEO investments in a review of related literature. We specifically explore: (A) high costs to access LEO; (B) launch challenges, especially those associated with secondary payloads; (C) price opacity associated with launch; (D) regulatory risks of space operations and investor perception problems of NASA; and finally (E) uncertain exit routes and financing risks for the sector. In short, though, the biggest barriers were cost, information uncertainty, regulatory risk, and uncertain exits.

---

24 David Livingston (now adjunct professor at the University of North Dakota in The School of Graduate Studies of Space Studies) used survey data (1988, 1996, 1998, and 1999) to gather practitioner and investor perspectives of space-related (dubbed "New Space") industries. The surveys reached a wide variety of audiences: aerospace executives, academia, VC firms (1996 and 1998 surveys), and commercial space entrepreneurs, space advocates, and representatives from space agencies such as NASA. Livingston noted that in spite of the diverse audience the surveys revealed common barriers to space commercialization: high costs of entering space, high insurance expenses, long development times, unfavorable government policies, overwhelming uncertainties, inexperienced space company management, and legal issues. The venture capitalists were particularly concerned with a lack of managerial experience in new space ventures, as well as high business and political risks, limited market size, and project costs. Given these risks, venture capitalists demanded returns in excess of 50 percent. See, David M. Livingston "The Obstacles to Financing New Space Industries." Mars Society (August 13, 1999); Condensed from a doctoral dissertation, Golden Gate University, San Francisco, California, 1999. Steve Jurvetson echoed these statements in reflecting on the space startups (launch vehicles, rail guns for fuel depots, space elevators, subcomponents for new propulsion systems) he met with over the roughly 10 year period to between 1995 and 2005 and noted similar barriers, namely, "…the amount of money required; the size of the market; [and] the timing and dependency on others, be it governments or other industry giants…to make your business model work." In fact, he stated that "conventional wisdom [held] that [the private space industry] wasn't a venture investment." See Steve "Small Sat 2014: Keynote Steve Jurvetson." YouTube video, 1:08:27. Posted by Small Sat Conference, August 18, 2014. *https://www.youtube.com/watch?v=qzudBqGyPTY#t=340* [7:30–8:18].

A. High Costs of Entry

We first look more closely into costs associated with LEO access, which have traditionally been viewed as among the largest obstacles for private players.[25] In response to a rising focus on launch costs among policy makers, space-related entrepreneurs, and satellite operators looking to commercialize space, one study undertook an in-depth analysis of payload costs from 1990 to 2000. To obtain the most accurate per-pound launch prices, the authors divided launch prices (or alternatively the average price for that launch vehicle) by the actual total mass of payloads, rather than payload capacities. Using this method, the authors found non-geosynchronous orbits (NGSO) payloads typically charged around $10,000 (constant year 2000 U.S. dollars) per pound in the late 1990s.[26]

Launch costs were exacerbated by the unknown but nontrivial chance of failure, which would not just destroy the product but cost the company its place in the launch sequence and send it back to the end of the line. Estimated in 1998 at roughly nine percent, LEO launch failure could instantly bankrupt a young competitor.[27] Even if the project got into space, space debris or component failure could render technologies inoperable.[28] Given these risks, insurance encompassing launch, delivery, and liability has historically reached 15 percent of vehicle, payload, service, and ground facility costs.[29]

To contextualize the capital constraints faced by LEO entrepreneurs, we look at the total costs for the three main commercial NGSO satellite systems in the late-1990s that together accounted for 86 percent of worldwide commercial non-geosynchronous orbit (NGSO) launches from 1997 to 1999: Iridium, Globalstar, and

---

25 A NASA-sponsored report on space commercialization noted, for example, that "commercial and government space market growth has been severely limited by decades of consistently high costs of space access." See, Hoyt Davidson, John Stone, and Ian Fichtenbaum. "Part 2: Support Alternatives Versus NASA Commercialization Priorities." In Supporting Commercial Space Development, New York, NY: Near Earth LLC, November 2010. This view was expressed by Alex Saltman, formerly the Executive Director of the Commercial Spaceflight Federation (a private spaceflight industry group founded in 2005), who noted, "The high cost of space launch has been the single biggest barrier to the broader exploration and development of space in the last forty years." See, Alex Saltman. "Commercial Spaceflight for Science and Exploration." Commercial Spaceflight Federation (April 2, 2014).

26 Authors used constant 2000 USD. These prices likely are not indicative of the per pound price for small payloads. See, Futron Corporation. Space Transportation Costs: Trends in Price Per Pound to Orbit 1990–2000, September 2002.

27 Ed Kyle. "Space Launch Report: 1998 Launch Log and Launch Vehicle/Site Statistics," *Spacelaunchreport.com* (1998).

28 Hoyt Davidson, John Stone, and Ian Fichtenbaum. "Part 1: Support Alternatives Versus Investor Risk Perceptions & Tolerances." In *Supporting Commercial Space Development*. New York, NY: Near Earth LLC, p. 36.

29 John Jurist, Sam Dinkin, and David Livingston. When Physics, Economics, and Reality Collide: The Challenge of Cheap Orbital Access. Colony Fund, 2007.

ORBCOMM.[30] These systems respectively cost an estimated $5 billion, $3 billion, and $500 million. Each of Iridium's 66 LEO satellites weighed 1,500 pounds (first launch occurred in 1997), and each of Globalstar's 48 LEO satellite constellation weighed 985 pounds.[31] ORBCOMM instead operated a 48-unit constellation of smaller 95-pound satellites. Iridium was originally developed and backed by telecommunications giant Motorola; Globalstar was backed by a consortium of 10 telecommunications companies; and ORBCOMM was originally founded by publicly traded Orbital Sciences Corporation and received additional backing from Canadian telecommunications firm Teleglobe Resources Industries and Technology Resources Industries, which controlled the largest cellular operator in Malaysia. The vast capital requirements of LEO satellites and deep pockets of industry incumbents generally rendered entrepreneurial competition infeasible.

B. "Secondary" Status of Miniature Satellites and Infrequent Launches

Critically, miniature satellites—a component of rising interest in the LEO sector and often deployed via ISS—have typically been restricted to "secondary payload" launch opportunities. As noted above, secondary payloads are stowed in the extra space remaining in launch vehicles after they have taken on their primary payloads.[32] The availability of secondary payload opportunities for miniature satellites has been further constrained by the limited number of launches available. In 2010, Jason Andrews and Jeff Cannon of Spaceflight Services, for example, noted that secondary payload opportunities have historically been "limited and sporadic" due to logistical issues, such as mismatched orbit destinations and a lack of available capacity.[33] While precise data on the average wait time for secondary payloads to access LEO are unclear, practitioners have reported launch lead times of "years."[34]

C. Price Opacity to Enter LEO

Even the true price of launching an item into LEO has historically been shrouded in mystery. In our interviews, one venture capitalist backing a miniature satellite company explained that actual costs to launch the firm's first satellite into LEO were

---

30 See, Federal Aviation Administration and the Commercial Space Transportation Advisory Committee. 2002 Commercial Space Transportation Forecasts. FAA and COMSTAC, May 2002, p. 30.

31 Numbers do not include spare satellites.

32 Shahed Aziz, Paul Gloyer, Joel Pedlikin, and Kimberly Kohlhepp. "Universal Small Payload Interface—an Assessment of U.S. Piggyback Launch Capability." Technical Session XI: Advanced Subsystems and Components II. SSC00-XI-3. Proceedings of the 14th Annual AIAA/USU Conference on Small Satellites. 2000.

33 Jason Andrews and Jeff Cannon. "Routine Scheduled Space Access for Secondary Payloads." Technical Session IX: From Earth To Orbit. SSC10-IX-8. Proceedings of the 24th Annual AIAA/USU Conference on Small Satellites. 2010, 1.

34 See, for instance, Peter M. Wegner, Jeff Ganley, and Joseph R. Maly. "EELV Secondary Payload Adapter (ESPA): Providing Increased Access to Space." Proceedings from IEEE Aerospace Conference. Big Sky, MT: IEEE, March 2001.

three times the projected estimates. Such opacity, unsurprisingly, has consistently dissuaded entrepreneurs from considering space as a medium for development and thus reduced any interest that venture capitalists might have had in investing there. In fact, Futron Corporation found that final negotiated launch costs varied greatly and were contingent on "…customer requirements, the existing supply of and demand for launch services, and any special provisions."[35] Without widely published figures on the costs of payloads, entrepreneurs, in the words of venture capitalist Steve Jurvetson, had to be or know of an "insider" to get the "real" launch price.[36] With the final negotiated prices a "black art of the insiders," would-be entrepreneurs could not answer the most basic questions about getting to space.[37]

D. Regulatory Barriers and NASA Perception Problem

Entrepreneurial efforts in the LEO sector have historically been further inhibited by regulatory uncertainty and compliance costs. Most notably, a 2010 report of the aerospace investment bank Near Earth found that the U.S. Department of State's International Traffic in Arms Regulations (ITAR)—which implements the Arms Export Control Act of 1976[38]—was a serious concern for the satellite manufacturing and launch services markets.[39] The authors specifically noted that ITAR has constrained fundraising in the market for low cost reliable access to space (primarily launch vehicles).

The way in which entrepreneurs viewed NASA has also impeded early stage investment in commercial space ventures. The report authors found that NASA's requirements were generally viewed as "unachievable" or "too costly" for the space industry and that NASA demotivated entrepreneurs through subsidization of competing systems.[40] In addition, NASA's sheer size overwhelmed small startups. Bill Claybaugh, previously Senior Director for human space systems at Orbital Sciences Corporation, explained, "[o]ne of the shocking things that all startups go through in dealing with the government is the day that they're holding some review and the company has four

---

35 Futron Corporation. Space Transportation Costs: Trends in Price Per Pound to Orbit 1990–2000, September 2002. The opaqueness of industry prices is also express in David Kestenbaum. "Spaceflight Is Getting Cheaper. But It's Still Not Cheap Enough." *NPR.org* (July 21, 2011).

36 "Small Sat 2014: Keynote Steve Jurvetson." YouTube video, 1:08:27. Posted by Small Sat Conference, August 18, 2014. *https://www.youtube.com/watch?v=qzudBqGyPTY#t=340* [12:59–14:00].

37 "Small Sat 2014: Keynote Steve Jurvetson." YouTube video, 1:08:27. Posted by Small Sat Conference, August 18, 2014. *https://www.youtube.com/watch?v=qzudBqGyPTY#t=340* [12:59–14:00].

38 Arms Export Control Act, 22 U.S.C. 39, § 2751 et seq.

39 Hoyt Davidson, John Stone, and Ian Fichtenbaum. "Part 2: Support Alternatives Versus NASA Commercialization Priorities." In Supporting Commercial Space Development. New York, NY: Near Earth LLC, November 2010, p. 8.

40 Hoyt Davidson, John Stone, and Ian Fichtenbaum. "Part 1: Support Alternatives Versus NASA Commercialization Priorities." In Supporting Commercial Space Development, New York, NY: Near Earth LLC, November 2010, p. 58.

people and the government shows up with fourteen ... Boeing and Lockheed know to send twenty even though the presentation only takes four because this is a contact sport. You need to be one-on-one with every one of those government guys to make sure they are getting what they want."[41]

E. Uncertain Exit Routes and Financing Risk

Finally, inchoate markets without any "shining star" benchmarks have historically been unattractive to venture capitalists. Because VC firms prefer to have an exit strategy in mind when they make an investment—and the first notable exit of a VC-backed space firm happened recently—few firms were interested in a sector that seemed so unlikely to provide gains to their investors.[42] The report authors further emphasized that for companies enabling low cost reliable access to space a "[c]ombination of large capital needs and high return expectations given the high risks involved makes it difficult to close business cases," especially given a "[h]istory of costly commercial failures."[43] In fact, each of the three major LEO satellite companies of the late 1990s (Iridium, Globalstar, and ORBCOMM) had filed for bankruptcy between 1999 and early 2002 due to commercially impractical business models and a lack of market demand.[44]

With respect to the ISS, the commercial laboratory subsector also faced a variety of distinct barriers that limited exit potential. Most notably, the report authors ascribe much of the limited LEO laboratory-related market development to a lack of industry awareness of the benefits of manufacturing in microgravity by pharmaceutical and materials companies.[45]

---

41 Ted O'Callahan. "Is There Profit in Outer Space?" *Yale Insights* (December, 2011).

42 This challenge was echoed by Stephen Fleming of the Space Angels Network. See, Jeff Foust. "Lawyers, Insurance, and Money: The Business Challenges of NewSpace." *Thespacereview.com* (March 26, 2007).

43 Hoyt Davidson, John Stone, and Ian Fichtenbaum. "Part 2: Support Alternatives Versus NASA Commercialization Priorities." In Supporting Commercial Space Development. New York, NY: Near Earth LLC, November 2010, 43–44.

44 See, Federal Aviation Administration and the Commercial Space Transportation Advisory Committee. 2002 Commercial Space Transportation Forecasts. FAA and COMSTAC, May 2002, p. 30. For example, one author wrote, " ... Globalstar's phones were quite large, worked only in open fields, had frequent dropped calls, and cost over $1000 ... its telephone calls cost $2–$3 per minute ... and its coverage areas were very limited. As a result, in 2000, it was reported that Globalstar actually had only a few thousand paying customers who typically used their phones for less than 30 minutes per month." The same author explained that in 1999 ORBCOMM's investors were " ... not prepared to absorb substantial ongoing losses associated with debt repayments, additional facilities, and operating expenses, as " ... many customers were simply not willing to purchase equipment and sign onto a brand new type of service until that service had been fully deployed and operating for some time." See, Roger Cochetti. *Mobile Satellite Communications Handbook*. Hoboken, NJ: Wiley, 2014.

45 Hoyt Davidson, John Stone, and Ian Fichtenbaum. "Part 2: Support Alternatives Versus NASA Commercialization Priorities." In Supporting Commercial Space Development, New York, NY: Near Earth LLC, November 2010, 85, 88.

A natural question is whether the maturation of the industry (and overlapping industries) has reduced skepticism among venture capitalists today.

The Current VC Perspective of and Emerging Opportunities in the LEO Sector

Despite these many challenges, venture capitalists have shown emerging interest in the LEO sector over the past five years. The sector's rising attraction is attributable to a variety of factors, including (A) lower launch costs and more transparent prices; (B) more frequent launch options and opportunities, especially for miniature satellites; and (C) the emergence of miniature satellites with commercial applications. While improving, complex and cumbersome federal regulations govern the sector, and it does not appear that venture capitalists have been attracted to the microgravity (ISS) manufacturing subsector. We explore each of these topics in more detail below.

A. The Beginnings of Reduced Launch Costs and More Transparent Prices

Given the barriers noted earlier, VC interest in LEO opportunities—such as they were—could only be expected after some sort of role-model company appeared. According to prominent LEO venture capitalist Steve Jurvetson of Draper Fisher Jurvetson (DFJ), that model emerged in 2002 with the founding of Space Exploration Technologies (SpaceX), a commercial launch company designed to lower the cost of entry to space.[46] SpaceX was founded by Elon Musk, who had co-founded the VC-backed online payment service provider PayPal in 1999. As of January 2015, SpaceX was valued at $12 billion.

Musk started SpaceX with $100 million of personal capital in 2002. NASA awarded SpaceX $278 million in seed funding under the Commercial Orbital Transportation Services (COTS) program to stimulate commercial transportation to the ISS.[47] By September 2008 and after three failed attempts, SpaceX's prototype launch vehicle (the Falcon 1), which was designed for smaller LEO payloads, became the first privately developed liquid-fueled rocket to orbit Earth. While SpaceX eventually abandoned the Falcon 1, the firm received $1.6 billion from NASA in 2008 for 12 ISS cargo resupply to the ISS. As of January 2015, SpaceX had completed 14 successful missions with its Falcon 9 vehicle, five of which were commercial resupply service missions to the ISS.[48] SpaceX's current version of the Falcon 9 (Falcon 9 v1.1) can

---

[46] "Small Sat 2014: Keynote Steve Jurvetson." YouTube video, 1:08. Posted by Small Sat Conference, August 18, 2014. *https://www.youtube.com/watch?v=qzudBqGyPTY#t=340* [8:00–13:00].

[47] Brian Berger. "SpaceX, Rocketplane Kistler Win NASA COTS Competition." *Space.com* (August 18, 2006).

[48] For a list of completed missions, see *http://www.spacex.com/missions*. For information regarding SpaceX's fifth resupply trip to the ISS, see, "SpaceX Launches Fifth Official Mission to Resupply the Space Station." SpaceX Blog (January 10, 2015). We do note that SpaceX was unable to deploy a secondary payload into the planned orbit on its October 2012 launch. See, "First Outing for SpaceX." *nytimes.com* (October 29, 2012).

lift payloads of almost 30,000 pounds (15 tons) to LEO and costs $61.2 million per launch, which experts suggest is the lowest figure in the industry.[49]

As of 2015, the company had made major strides in validating a "game-changing" low-cost launch technology critical to the LEO sector. Jurvetson compared the impact of reductions in launch costs to the impact of fiber optics for internet services companies, explaining that just as fiber optics encouraged the emergence of new business models such as Hotmail and Skype, so too [would] reduced launch costs provide a "ray of hope" to spur innovative business models from space-focused entrepreneurs.[50]

Although the $61.2 million price tag for a 2016 Falcon 9 launch is likely to be a high barrier for many entrepreneurs, SpaceX also offers secondary launch services aboard the Falcon 9 for miniature satellites, which helps to support this emerging subsector (discussed in more detail below). According to Dustin Doud, et al., SpaceX utilizes secondary payloads if the primary payload fills less than 80 percent capacity.[51]

Another major advancement in launch costs and price transparency for miniature satellites comes from "specialist" small satellite coordinators, which leverage international partnerships with launch vehicles to create a pipeline of launches at commercial prices. Space logistics company Spaceflight Industries, for example, offers baseline price quotes for the launch of miniature satellites based on the weight/size of the payload and orbit destination. CubeSats (5 kg) can access LEO for roughly $295,000 and 100 kg microsatellites for under $4 million.[52] Companies such as Spaceflight not only reduce launch costs, but also enable entrepreneurs to more accurately gauge budgetary requirements with relative ease.

B. Infrequent Launches Remain an Issue, But Progress is Being Made

While lower launch costs have helped spur entrepreneurial efforts in the LEO sector, launch frequencies remain suboptimal. For example, research suggests substantial delays for launches on which satellites seek rideshare opportunities.[53] As evidence, a SpaceWorks report found that that "[t]here was little excess capacity for nano/micro-

---

49 Justin Bachman. "To Reuse Rockets, SpaceX Needs to Stick a Tricky Landing." *Bloomberg.com* (January 5, 2015).

50 "Small Sat 2014: Keynote Steve Jurvetson." YouTube video, 1:08:27. Posted by Small Sat Conference, August 18, 2014 *https://www.youtube.com/watch?v=qzudBqGyPTY#t=340* [12:00–13:00].

51 Dustin Doud, Brian Bjelde, Christian Melbostad, and Lauren Dreyer. "Secondary Launch Services and Payload Hosting Aboard the Falcon and Dragon Product Lines." Technical Session V: Getting There. SSC12-V-3. Proceedings of the 26th Annual AIAA/USU Conference on Small Satellites. 2012.

52 See *http://spaceflightservices.com/pricing-plans*, accessed January 15, 2015.

53 Adam Snow, Elizabeth Buchen, and John R. Olds. Global Launch Vehicle Market Assessment: A Study of Launch Services for Nano/Microsatellites in 2013. Atlanta: SpaceWorks, July 2014, 11. Smallsat dedicated launch company Rocket Lab (discussed below) noted a lack of "responsiveness" in launch opportunities for LEO small satellites. See, Peter Beck, Interview by Spacevidcast. "The Electron Rocket – 7.24." YouTube video, 48:53. Posted by Spacevidcast, August 10, 2014. *https://www.youtube.com/watch?v=tkmrZVDmio4* [25:00–27:00].

satellites (1–50 kg) on 2013 launches [to LEO] given vehicle integration limitations" and concludes that "… the current supply of launch vehicles will not sufficiently serve future nano/microsatellite market demand."[54]

In response to this challenge, as well as a spike in demand for miniature satellite launches (discussed below), the LEO sector has seen two key advances: (i) miniature satellite launch brokerage and "concierge-type" deployment via ISS, and (ii) the development of low-cost dedicated small satellite launches.

i. Miniature Satellite Launch Brokerage and "Concierge-Type" Deployment Via ISS

We identified a number of companies that were expanding launch opportunities to LEO for miniature satellites. We examine two such companies in more detail: Spaceflight Industries (referenced above) and NanoRacks.

One venture capitalist explained that Spaceflight works with a broad network of commercially accessible launch operators worldwide (such as SpaceX, Orbital Sciences, Roscosmos, and Virgin Galactic) and has emerged as "the principal broker for secondary market launches." Spaceflight provides the manifests for, certifies, and integrates secondary payloads on launch transportation vehicles, such as those from SpaceX and Orbital Sciences. Spaceflight also facilitates launch opportunities with integration services, and provides ITAR expertise and technical assistance to minimize the risk of regulatory snafus or malfunction. Spaceflight documentation (as of March 2013) suggests that it can provide launches between 18 and 24 months after contract signing.[55]

Another company that facilitates launches for miniature satellites is NanoRacks. NanoRacks is a privately funded, VC-backed startup founded in 2009 that entered into a Space Act Agreement with NASA in September 2009 to develop proprietary commercial research facilities on the ISS. Of particular interest to venture capitalists, the NanoRacks Smallsat Deployment Program utilizes launch vehicles to the ISS for small satellite deployment. Importantly, NanoRacks offers relatively frequent launch opportunities given consistent cargo missions servicing the ISS. NanoRacks also coordinates "… payload integration, payload design and development, and interfacing with NASA and foreign space agencies."[56] The firm's documentation suggests varied turnaround times from contract signing to launch, from nine months[57] to 12–14 months.[58]

---

54 Adam Snow, Elizabeth Buchen, and John R. Olds. Global Launch Vehicle Market Assessment: A Study of Launch Services for Nano/Microsatellites in 2013. Atlanta: SpaceWorks, July 2014, 10–11.

55 Spaceflight, Inc. Secondary Payload Users Guide: Spaceflight Inc., 2013.

56 For overview of the NanoRacks Smallsat Deployment Program, see, *http://nanoracks.com/products/smallsat-deployment*.

57 See NanoRacks slide deck, at *http://www.nasa.gov/sites/default/files/accelerating_innovation_through_microgravity_research.pdf*.

58 NanoRacks. *NanoRacks CubeSat Deployer (NRCSD) Interface Control Document.* Houston: Nano-Racks, 2013.

As of October 2014, NanoRacks had deployed over 40 CubeSats.[59] While NanoRacks' deployment is mostly limited to the "nano" end of small satellites and has encountered technical difficulties resulting in suspended missions, the program may spur entrepreneurial and VC activity in the LEO sector by offering more affordable, more frequent, and more reliable—even if not perfect—access to the domain.

ii. Low-Cost Miniature Satellite Dedicated Launches

To date, Spaceflight and NanoRacks have streamlined the launch process for miniature satellites via secondary payloads. The constraints inherent to flying secondary remain, however. For example, deployment via NanoRacks necessitates launch at the orbit of the ISS, which several of our interviewees explained may be suboptimal for some payloads.[60] In addition, launch and integration schedules are still in the hands of the primary payload operators, which adds additional layers of uncertainty. As a result, startups are building low-cost launchers dedicated to miniature payloads.

A new low-cost, miniature satellite-dedicated launcher is being developed by Khosla Ventures-backed Rocket Lab. Rocket Lab's Electron launch vehicle is currently planned to deliver payloads up to 110 kg to LEO. According to Rocket Lab CEO Peter Beck, the company hopes to provide a solution to the two key barriers to space commercialization: launch cost and "responsiveness." First, Rocket Lab aims to reduce the total amount of capital that must be raised to enter space. Beck explained in a Spacevidcast interview: "For the same price [roughly $4.9 million for approximately 110 kg] of a rideshare you get to where you want to go, when you want to go."[61] He noted, however, that reasonable cost is largely irrelevant without timely, "responsive" launches. As a result, the company is aiming for 100 launches per year.[62]

Rocket Lab is not alone in its mission to provide high-frequency, low-cost small satellite-dedicated launches. The first launch of Virgin Galactic's miniature satellite-dedicated vehicle LauncherOne, for example, is scheduled in 2016 with customers such as Skybox Imaging, GeoOptics, and Planetary Resources.[63] If proven efficient, the vehicles may collectively reduce entrepreneurial barriers and spur VC interest in the sector. We note, however, that while the venture capitalists we interviewed generally

---

59 See "NanoRacks CubeSat Deployer (NRCSD) on the ISS." YouTube video, 1:31. Posted by Nano-Racks, October 1, 2014. *https://www.youtube.com/watch?v=AdtiVFwlXdw&t=19*.

60 Note, however, that Altius Space Machines' HatchBasket smallsat deployer carrier is working with NanoRacks to solve this problem by enabling deployment of spacecraft at altitudes higher than the ISS. See, Jonathan Goff, "HatchBasket: Genesis of a Concept." *Altius-Space.com* (August 4, 2014).

61 Peter Beck, Interview by Spacevidcast. "The Electron Rocket–7.24." YouTube video, 48:53. Posted by Spacevidcast, August 10, 2014. *https://www.youtube.com/watch?v=tkmrZVDmio4* [25:00–27:00].

62 See Rocket Lab website, at *http://www.rocketlabusa.com/our-mission*.

63 Jeff Foust. "Virgin Galactic's LauncherOne on Schedule for 2016 First Launch." *Spacenews.com* (March 16, 2015).

saw such launch vehicles as a step in the right direction, opinion was divided as to whether rideshares will likely remain dominant, at least in the near future.

C. Technological Advancement and "Big Data" Applications for Miniature Satellites

This increase in secondary and miniature satellite dedicated launch opportunities has largely been a response to the emergence of commercial nanosatellite and microsatellite startups. Recent projections suggest strong growth among nano/microsatellites (1–50 kg), stemming in large part from demand in the commercial sector. Whereas historically (2009–2013) the commercial sector was responsible for roughly 8 percent of nano/microsatellites globally, this figure is expected to spike to 56 percent for 2014–2016.[64] In fact, 2013 saw a 330 percent increase in attempted nanosatellite (1–10 kg) deliveries, compared to 2012.[65]

To further explore the changes contributing to recent VC interest in LEO, we explore the histories of two VC-backed LEO satellite companies: Planet Labs and Skybox.

i. Miniature Satellite Case Study: Planet Labs

Company Background: Planet Labs is a San Francisco-based startup founded in 2010 by three former NASA employees who sought to create a constellation of Earth-imaging nanosatellites (called "Doves") with data distribution applicable to commercial enterprises and humanitarian purposes. The company deploys its fleet of CubeSats, which typically weigh between five and six kilograms each, to collect global imagery data with an optical resolution of three to five meters at a much faster pace (daily) than that of traditional satellites.[66] These lightweight, shoebox-sized satellites travel as secondary payloads on larger launch vehicles.[67] In February 2014, Planet Labs deployed its "Flock 1" 28-satellite fleet from the ISS.[68] Subsequent successful deployments (as of January 2015) have included 22 satellites as part of the 28-satellite "Flock

---

64 Dominic DePasquale and Elizabeth Buchen. 2014 Nano/Microsatellite Market Assessment. Atlanta: SpaceWorks Enterprises, 2014, p. 9.

65 Dominic DePasquale and Elizabeth Buchen. 2014 Nano/Microsatellite Market Assessment. Atlanta: SpaceWorks Enterprises, 2014.

66 Kirk Woellert, Pascale Ehrenfreund, Antonio J. Ricco, and Henry Hertzfeld. "Cubesats: Cost-Effective Science and Technology Platforms for Emerging and Developing Nations." *Advances in Space Research* 47, no. 4 (February 15, 2011): 663–684.

67 "Flock 1 – Planet Labs Earth Observation Satellites." *Spaceflight101.com* (January 5, 2015). James Mason of Planet Labs noted that because the firm's satellites fly as secondary payloads, Planet Labs has no authority over final orbit parameters and sometimes accepts less-than-ideal orbit regions. See, James Mason. "Keeping Space Clean: Responsible Satellite Fleet Operations." Planet Labs Blog (October 16, 2014).

68 Mike Wall. "First 'Cubesats' in Record-Breaking Fleet Launched from Space Station." *Space.com* (February 11, 2014). See also, "Flock 1 – Planet Labs Earth Observation Satellites." *Spaceflight101.com* (January 5, 2015).

1b,"[69] an 11-satellite "Flock 1c," and a two-satellite "Flock-1d Prime," which came after a fleet of 26 "Flock 1d" satellites were destroyed by rocket failure.[70]

Planet Labs works in close partnership with NanoRacks and thus primarily (with the exception of Flock 1c) deploys satellites from the ISS. An issue associated with ISS deployment is a short orbital lifetime that inhibits commercial viability. As a result, Planet Labs' satellites have a lifespan of just one to three years, notably shorter than larger communications satellites.[71] The "replenishment model," however, allows for more frequent technology updates for each subsequent flock.[72]

Planet Labs distinguishes itself from traditional satellite companies in both its manufacturing methods and imaging rates. Planet Labs leverages consumer technology (e.g., mobile phones, laptops) to manufacture its satellites at a fraction of typical costs.[73] In a discussion of Planet Labs' Commercial, Off-the-Shelf (confusingly also bearing the acronym COTS associated with NASA's Commercial Orbital Transportation Services program) investment strategy, Planet Labs explained "…we look a lot more like a cell phone manufacturer than we do an aerospace company."[74] In line with an observation by Steve Jurvetson that space-related companies were integrating software economics into operations,[75] they noted that the company follows the software adage to "release early and often" and contrasted Planet Labs' 8–12 week iteration from design to manufacture of its Dove satellites with the years or decades for other devices in the aerospace industry.[76] The company leverage an "agile development" strategy, adopted from the software industry, achieving 10 builds in

---

69 For more detailed information, see *http://space.skyrocket.de/doc_sdat/flock-1.htm*.

70 For detailed information on Planet Labs' Flock 1 launches, see "Flock 1 – Planet Labs Earth Observation Satellites." *Spaceflight101.com* (January 5, 2015).

71 Ryan Lawler. "After Sending 2 'Doves' into Orbit, Planet Labs Prepares Largest Satellite Constellation for Launch." *Techcrunch.com* (November 26, 2013). For more detail as to the specifications of Planet Labs' satellites, see Christopher R. Boshuizen, James Mason, Pete Klupar, and Shannon Spanhake. "Results from the Planet Labs Flock Constellation." Technical Session I: Private Endeavors. SSC14-I-1. Proceedings of the 28th Annual AIAA/USU Conference on Small Satellites. 2014.

72 Debra Werner. "Profile: Chris Boshuizen, Chief Technology Officer, Planet Labs Inc." *Spacenews.com* (January 27, 2014).

73 Quentin Hardy and Nick Bilton. "Start-Ups Aim to Conquer Space Market." *nytimes.com* (March 16, 2014). One article suggested that cost is "substantially less than $100,000." See, Rakesh Sharma. "All Set for Take-Off: Silicon Valley Startups Redefine Space Imaging Market." *Forbes.com* (February 26, 2014).

74 Christopher R. Boshuizen, James Mason, Pete Klupar, and Shannon Spanhake. "Results from the Planet Labs Flock Constellation." Technical Session I: Private Endeavors. SSC14-I-1. Proceedings of the 28th Annual AIAA/USU Conference on Small Satellites. 2014, 2.

75 Jeff Foust. "The Silicon Valley of Space could be Silicon Valley." *Thespacereview.com* (July 29, 2013).

76 Christopher R. Boshuizen, James Mason, Pete Klupar, and Shannon Spanhake. "Results from the Planet Labs Flock Constellation." Technical Session I: Private Endeavors. SSC14-I-1. Proceedings of the 28th Annual AIAA/USU Conference on Small Satellites. 2014, 5.

36 months. Their approach is akin to "beta-testing" in the software industry, as the engineers release early and iterate often.

VC Investments: As of January 2015, Planet Labs had raised more than $135 million in entire funding along with $25 million in debt. The first investment, a $13.1 million Series A round that closed in June 2013, came just a few months after Planet Labs successfully deployed two demonstration satellites to validate key technologies, such as the resolution of its imagery data.[77] Backers were a consortium of VC firms, including DFJ, Capricorn, O'Reilly AlphaTech Ventures, Founders Fund, Innovation Endeavors, Data Collective, and First Round Capital. In November 2013, the company launched two additional demonstration satellites and, in the following month, secured $52 million in Series B funding, led by Russian billionaire Yuri Milner and with additional investment from existing investors such as DFJ. In January 2015, the company secured an additional $70 million in Series C funding, which was led by San Francisco-based VC fund Data Collective. As part of this round, Planet Labs also announced a $25 million debt facility from Western Technology Investment.

Investment Thesis: The VC groups backing Planet Labs have expressed excitement about LEO and, more specifically, the miniature satellite industry. Steve Jurvetson of DFJ affirmed, "We're seeing unprecedented innovation in the space industry, starting with SpaceX reducing the cost of access, and now with Planet Labs revolutionizing the satellite segment."[78] Jurvetson noted two specific developments in the satellite industry that were spawning high-potential startups: the standardization of secondary payloads and specialized launch services for small satellites.[79]

Another industry trend attractive to venture capitalists is the adoption of low-cost technologies used in other industries to speed development times, as opposed to the capital-intensive business model of traditional space-related companies. This trend is abundantly clear with Planet Labs, which utilizes, for example, facilities built by the automobile industry to perform heat transfer analysis and electronic design tools and testing houses from consumer electronics.[80] In broad terms, they found that advances

---

[77] For example, its Dove-1 satellite demonstrated its forestry application potential by capturing detailed imagery of a forest in Portland, Oregon. See Debra Werner. "Planet Labs Unveils Plan to Launch 28 Nanosats on Antares' 1st Cargo Run." *Spacenews.com* (June 26, 2013).

[78] See Leena Rao. "Planet Labs Raises $13M from DFJ, OATV, Founders Fund to Build the World's Largest Fleet of Earth-Imaging Satellites." *Techcrunch.com* (June 25, 2013).

[79] "SN Profile: Steve Jurvetson, Managing Director, Draper Fisher Jurvetson (DFJ)." *Spacenews.com* (July 29, 2013).

[80] Christopher R. Boshuizen, James Mason, Pete Klupar, and Shannon Spanhake. "Results from the Planet Labs Flock Constellation." Technical Session I: Private Endeavors. SSC14-I-1. Proceedings of the 28th Annual AIAA/USU Conference on Small Satellites. 2014.

in commercial technology have led to advances in CubeSats' power and reliability, which historically have been primary limiting factors.[81]

Finally, the investors see substantial commercial value in the data applications of Planet Labs' satellites. In fact, its most recent round of fundraising was led by a VC firm (Data Collective) that focuses specifically on "big data" startups. Tim O'Reilly of O'Reilly AlphaTech Ventures, which has invested in two rounds, further affirmed the transformative nature of "regularly and frequently" updated Earth imagery data.[82] As of early-February 2015, the company disclosed three clients:

- **Woolpert:** Design, geospatial, and infrastructure management firm (October 2014).

- **Geoplex:** Mapping and geographic information systems (GIS) firm (November 2014). Major applications include emergency management, agriculture, and forestry.[83]

- **Wilbur-Ellis:** Agricultural and industrial products marketer and distributor (February 2015).[84]

VC investors remain aware of substantial risk inherent to the LEO sector, as clearly demonstrated in the October 2014 explosion of the Antares rocket carrying 26 satellites.[85] The company anticipates such mishaps, however, and spreads launch risk across multiple vehicles from multiple vendors, such that the Antares explosion was "not catastrophic" to the firm.[86]

ii. Miniature Satellite Case Study: Skybox

Now, we turn to another LEO player. Skybox is developing larger microsatellites, but in many ways it operates under the same philosophy as Planet Labs. We note, however, that Skybox has, to date, not deployed its satellites via ISS.

Company Background: Skybox Imaging is a Mountain View, California-based firm founded in 2009 that builds high-resolution earth observation microsatellites. Similar to Planet Labs, the firm uses its satellites as a platform to generate data applicable to such industries as agriculture, oil and gas, finance, and mining, as well as a variety of humanitarian efforts. As explained by co-founder Dan Berkenstock, "What's exciting to [Skybox] … [is] not just the picture of the parking lot, but an answer for how

---

81 Christopher R. Boshuizen, James Mason, Pete Klupar, and Shannon Spanhake. "Results from the Planet Labs Flock Constellation." Technical Session I: Private Endeavors. SSC14-I-1. Proceedings of the 28th Annual AIAA/USU Conference on Small Satellites. 2014, 3.

82 "Planet Labs Reveals First Images from Space; Announces Plans to Launch Fleet of Satellites to Understand the Changing Planet." *BusinessWire.com* (June 23, 2013).

83 Beau Jarvis. "Geoplex and Planet Labs Partner to Supply Australia and New Zealand with Satellite Imagery." Planet Labs Blog (November 20, 2014).

84 Josh Alban. "Planet Labs Strikes Agreement with Wilbur-Ellis to Enhance AgVerdict® Data Tool." Planet Labs Blog (February 2, 2015).

85 Will Marshall. "Space is Hard: Antares Rocket Failure." Planet Labs Blog (October 28, 2014).

86 Will Marshall. "Space is Hard: Antares Rocket Failure." Planet Labs Blog (October 28, 2014).

many cars are within it and how...that [number] compare[s] to what happened last week, last month, last year."[87]

Skybox aims to build an LEO satellite constellation to capture imagery and collect data of a single spot on a frequent basis (five to seven times per day).[88] Skybox's first satellite—dubbed the SkySat 1—was launched aboard the Russian Dnepr rocket in November 2013. The company subsequently launched SkySat 2 in July 2014 aboard the Soyuz-2/Fregat rocket from Kazakhstan. Each of these satellites weighed roughly 100 kg—substantially larger than Planet Labs' 5–6 kg satellites—but was equipped with sub-meter resolution imagery (relative to three-to-five meters for Planet Labs), as well as high definition video capabilities.[89] Working toward a 24-satellite constellation, Skybox partnered with Space Systems/Loral (SSL) in February 2014 to build 13 additional 120 kg satellites that would capture images of any point on Earth three times per day.[90] In the same month, Skybox signed a contract with American launch company Orbital Sciences Corporation to launch six of the SSL manufactured devices aboard the Minotaur-C rocket from California, which will be equipped with a specially designed SkySat satellite dispenser.[91]

Skybox builds its low-cost satellites with COTS technologies. Ethan Kurzweil, partner at Bessemer Venture Partners and board observer, explained, "[i]ncumbents...launch these massive satellites that cost 10 years to make. By the time you launch, you're three cycles behind.... By taking off-the-shelf parts [and] putting them together in a rapid design and test methodology, you shorten up those cycles."[92] Using designs based on COTS components such as oscillators and circuit boards,[93] one source suggested that Skybox's first satellite (SkySat-1) cost roughly $2 million to $5 million to build.[94]

VC Investments (Pre-exit): Skybox raised a total of $94 million prior to its acquisition by Google in August 2014. Skybox received Series A financing of $3 million from Khosla Ventures in July 2009. With this funding, the founders moved operations out

---

87 "Stanford Seminar—Dan Berkenstock, Julian Mann, John Fenwick, and Ching-Yu Hu." YouTube video, 57:49. Posted by stanfordonline, March 22, 2013. *https://youtu.be/i-1dlUS3rEo?t=270* [4:31–5:15].

88 Andreas Jelinek. "Entering the Final Stretch for the European Space Imaging & Skybox Imaging High-Res Challenge." Skybox Imaging Blog (June 24, 2014).

89 "SkySat-1 First Light." Skybox Imaging Blog (December 11, 2013).

90 Space Systems/Loral (SSL). "Skybox Imaging Selects SSL to Build 13 Low Earth Orbit Imaging Satellites." (February 10, 2014).

91 Peter B. de Selding. "Skybox Taps Orbital Sciences for 2015 Minotaur Launch." *Spacenews.com* (February 20, 2014).

92 Liz Gannes and James Temple. "Everything You Need to Know about Skybox, Google's Big Satellite Play." *Recode.net* (June 11, 2014).

93 Debra Werner. "Skybox Imaging's Hopes High as Launch of First Satellites Draws Near." *Spacenews.com* (October 14, 2013).

94 Liz Gannes and James Temple. "Everything You Need to Know about Skybox, Google's Big Satellite Play." *Recode.Net* (June 11, 2014).

of co-founder John Fenwick's living room and into an office in Palo Alto and began to expand the team. One year later—after becoming the fifth U.S. company to receive a NOAA license to operate a high-resolution satellite—the firm closed Series B financing of $18 million from investors including Khosla Ventures and Bessemer Venture Partners. In April 2012, Skybox closed a $70 million Series C round to support the launches of SkySat-1 and SkySat-2 and expand strategic alliances. Finally, the company received an additional $3 million investment post Series C.

Investment Thesis: Skybox's initial fundraising efforts were marked by extreme skepticism among venture capitalists. According to Berkenstock, the founders had pitched Skybox over 50 times in the previous 6.5 months before closing its Series A financing.[95] In 2009, the Skybox team was introduced to Pierre Lamond of Khosla Ventures by their Stanford professor Mark Leslie, a former VC-backed CEO. With fewer than 30 global miniature satellite (1–50 kg) launches in 2009 from the technology, its commercial value was uncertain.[96] One venture capitalist explained that most other venture capitalists at the time were extremely skeptical of the investment.

By 2010, however, market viability was becoming increasingly clear. At the time Skybox closed its Series B round, Berkenstock spoke of a disequilibrium in the supply and demand equation of Earth observation satellites, "Commercial customers, including those in the oil-and-gas, mining, financial trading, agriculture and forestry industries, are hungry for more timely, high-resolution images for space. There simply aren't enough satellites out there…to provide the kind of daily, or even hourly updates that companies need."[97] While Skybox raised $18 million in July 2010, investors were still hesitant about the technology. David Cowan, who regarded Skybox as a "naïve pipe dream" in 2009, stated that he "built up the nerve" to invest in Series B.[98] Beyond technology questions, venture capitalists explained that another key constraint was the limited number of launch vehicles on which Skybox satellites could ride.

Nonetheless, of critical importance was the rising "big data" market in which Skybox (like Planet Labs) had a unique position. A March 2012 report by market intelligence firm IDC projected that the "Big Data technology and services market" would balloon from $3.2 billion in 2010 to $16.9 billion in 2015, which would represent a compound annual growth rate (CAGR) roughly seven times that of the overall

---

95 "Stanford Seminar—Dan Berkenstock, Julian Mann, John Fenwick, and Ching-Yu Hu." YouTube video, 57:49. Posted by stanfordonline, March 22, 2013. *https://youtu.be/i-1dlUS3rEo?t=706* [11:40–13:08].

96 Dominic DePasquale and John Bradford. 2013 Nano/Microsatellite Market Assessment. Atlanta: SpaceWorks Enterprises, February 2013.

97 "Skybox Imaging Raises $18M in Series B Financing." Skybox Imaging Blog (July 26, 2010).

98 Seth Fiegerman. "How 4 Students Got a Satellite Startup Off the Ground and $500 Million from Google." *Mashable.com* (June 11, 2014).

information and communication technology (ICT) market.[99] In particular, IDC projected the Big Data services segment to increase at CAGR 39.5 percent from $1.2 billion to $6.5 billion.[100] As a result, VC firms saw major potential in the industry. IDC found that VC funding in the big data and analytics software market increased from $155 million in 2009 to $726 million in 2011.[101]

In our interview, one venture capitalist further emphasized that VC backing, rather than NASA backing, could itself actually remove many of the historical impediments to the sector. He explained that VC backing "removed rules" in the domain, as ventures could bypass restrictions and liabilities stemming from government risk aversion. "Every time you build redundancy into a system, you add weight and cost."

The Exit: A turning point came in November 2013 when Skybox launched its SkySat-1 aboard the Dnepr rocket. The following month Skybox validated its sub-meter resolution technology, as it released several high-resolution images and a high-definition video. As noted by one venture capitalist, "[Once Skybox] sent down the first pretty picture, VCs changed their minds [about the company]." One board member also noted that post-launch, "…all sorts of suitors came out of the woodwork."[102] In February 2014, Skybox secured a launch agreement for six follow-on satellites (to be manufactured by Space Systems/Loral) in late 2015.

In August 2014, Google acquired Skybox for $500 million. While the acquisition would allow Google to improve its maps services, the primary applications were much broader: bringing satellite-based internet access to developing countries and capitalizing on data that could be extracted from the satellites.

Regardless of eventual applications, Skybox's acquisition closed the VC circle by demonstrating an exit for a LEO company. As one venture capitalist explained, "With Skybox's sale to Google and SpaceX [valued at $10 billion in January 2015], there were two winners in space and you could say that space [became] open for business."

Skybox and Planet Labs represent a new commercial model for satellite imaging. Each is in essence a big data company employing relatively low-cost miniature satellite technology to collect imagery data with high-value commercial applications. The revolution in such satellite technology is largely attributable to major advancements in the power and size of COTS hardware. The "softwarization" of R&D attitudes—e.g., the Agile principle of "iterate often"—has further driven down costs and development times, while allowing for greater risk and innovation. Relatedly, the development and integration of big data software that transforms sensory output into valuable commercial insights creates exit options for venture capitalists.

---

99 International Data Corporation. Worldwide Big Data Technology and Services 2012–2015 Forecast. IDC, March 2012.

100 International Data Corporation. Worldwide Big Data Technology and Services 2012–2015 Forecast. IDC, March 2012.

101 "VC Funding Trends in Big Data (IDC Report)." *Experfy.com* (June 11, 2014).

102 Seth Fiegerman. "How 4 Students Got a Satellite Startup Off the Ground and $500 Million from Google." *Mashable.com* (June 11, 2014).

D. Federal Regulations Remain a Hurdle but Are Improving

While miniature satellite companies like Planet Labs and Skybox have penetrated the LEO market, federal regulation has nonetheless remained problematic. In November 2011, Julian Mann, VP of Product Development & Research and co-founder of Skybox, alluded to major obstacles in adherence to Federal Communications Commission (FCC) requirements. In particular, he noted that FCC licenses for Earth observation satellites have extremely high capital requirements. Mann explained that ITAR has inflated manufacturing time and costs due to the requirement that every component be manufactured by ITAR-certified machine shops, and it has diminished domestic competitiveness by inhibiting collaboration with international developers who generally have client bases outside of the United States.[103]

We note, however, that the ITAR was since amended, and as of November 2014 most commercial satellites have become subject to the less onerous Bureau of Industry and Security's Export Administration Regulations (EAR).[104] In talks with a lawyer who specializes in regulatory issues for LEO miniature satellite firms, we found that while this change in regulation will ultimately relieve LEO companies like Skybox of much of the time and money devoted to ITAR compliance, it is not a "game-changing" law shift for the entrepreneurial miniature satellite community. This is because the shift in regulation does not constitute "decontrol," but rather a change in applicable regulations to a sometimes more complex regime. The shift in regulation does, however, allow commercial satellite companies to utilize License Exception Strategic Trade Authorization (STA) with respect to exports of to a number of countries in developed markets, such as Western Europe, Australia, New Zealand, Canada, and Japan, which he suggested will substantially reduce the administrative burden. In addition, "concierge-style" secondary payload brokers (such as Spaceflight) streamline compliance with federal regulations. Venture capitalists nonetheless continue to find that complicated federal regulations impose compliance costs and additional time to properly arrange export-related transactions.

We also note major concerns with respect to intellectual property (IP) and commercial users' willingness to conduct applied research. In the NASA Office of the Inspector General's Audit Report of September 2014, the authors explained that "[u]nder the 2011 cooperative agreement with NASA, CASIS and its partners must transfer patent licenses and data rights related to federally supported research conducted on Station to the Government... [which includes] discoveries funded by the

---

103 See Julian Mann's written statement from Creating and Growing New Businesses: Fostering U.S. Innovation. 112 Congress, First sess., November 2, 2011. For detailed information on FCC requirements, see, Michael Swartwout. Secondary Payloads in 2014: Assessing the Numbers. 2014 IEEE Aerospace Conference, 2014. See also, *http://www.hallikainen.com/FccRules/2014/25/165/index.php*.

104 "State and Commerce Departments Publish Interim Final Rules Implementing Satellite Export Control Reform." Covington & Burling (May 23, 2014).

Federal government via NASA's annual $15 million award."[105] The authors explicitly noted that such provisions dissuaded pharmaceutical and consumer product companies from working with the ISS. While amendments were made in 2012 to give private researchers "more control" of their research, they failed to assuage commercial users' concerns (according to CASIS). Finally, in 2013 the NASA Advisory Council recommended a legislative amendment to ensure that users (both large and small entities) retain all IP rights. The authors stated that as of July 2014 the proposal had not been implemented into congressional legislation.[106]

E. VC Interest in Non-Satellite-Related LEO Activity is Slim

Although venture capitalists have shown interest in launch vehicles and miniature satellites, they have expressed little interest in companies using microgravity in the biotechnology, pharmaceutical, and semiconductor industries. In our talks with venture capitalists involved in the sector, none mentioned any current interest in LEO pharmaceutical/biotechnology manufacturing. One company has received VC funding however.

ACME Advanced Materials (A2M) is an Albuquerque-based company established in 2014 that produces semiconductor materials in microgravity conditions.[107] While A2M has disclosed negligible operational detail to date, the firm announced commercialization of its "process to produce large quantities of low loss, electrically defect free (EDF) Silicon Carbide (SiC) wafers in a microgravity environment."[108] The commercial value of SiC lies primarily in applications to power electronic devices, as the material properties of SiC are more energy efficient relative to standard silicon (Si) wafers.[109] A2M received "seven-figure"[110] seed funding from New Mexico-based Cottonwood Technology Fund and Canada-based Pangaea Ventures. Cottonwood Technology Fund is a seed and early stage technology commercialization fund investing primarily in chemistry/material sciences, photonics, biosciences, and new energy related businesses.[111]

---

105 National Aeronautics and Space Administration. *Extending the Operational Life of the International Space Station Until 2024*. Washington, DC: NASA Office of Inspector, 2014, 132.

106 National Aeronautics and Space Administration. *Extending the Operational Life of the International Space Station Until 2024*. Washington, DC: NASA Office of Inspector, 2014.

107 A2M started as Medusa Space (founded in 2012). After an investor failed to provide promised funds, Medusa Space was reestablished as Masterson Industries. Once Masterson achieved proof of concept and received funding from new investors, the firm become A2M. See, Keith Cowing. "Playing 20 Questions with A Microgravity Company." *Nasawatch.com* (December 9, 2014).

108 "ACME Advanced Materials, Inc Announces First Commercial Production of 4" SiC Wafers in Microgravity." *PRWeb.com* (September 9, 2014).

109 F. Roccaforte, F. Giannazzo, and V. Raineri. "Nanoscale Transport Properties at Silicon Carbide Interfaces." *Journal of Physics D: Applied Physics* 43, no. 22 (2010): 223001.

110 Kevin Robinson-Avila. "Made in Space." *Albuquerque Journal* (October 13, 2014).

111 "ACME Advanced Materials, Inc. Announces First Commercial Production of 4" SiC Wafers in Microgravity." *PRWeb.com* (September 9, 2014).

Other than this investment, however, we have found VC interest in this niche to be thus far nonexistent. We have been unable to find any VC deals other than A2M that directly fall into the "microgravity-enabled" subsector, although there may be angel-funded deals that do not reach our databases.[112] Still, A2M President Glover asserted in late 2014 that while most commercial space-based investments are currently in the area of launch facilitation, "…the real bonanza will come from the materials and products that are manufacturable in microgravity."[113]

We note, however, a major "red flag" with respect to microgravity manufacturing entrepreneurship: there is currently a very limited supply of investable start-up companies in this nascent sector. In other words, microgravity manufacturing itself is in the development/seed stage, as it has no track record of market validation and no notable "role model" companies against which venture capitalists can benchmark. The establishment, however, of a demonstrated commercial success in this area could—like with the case of SpaceX and SkyBox—generate increased interest.

Section 3. A Quantitative Look at Venture Capital Activity in LEO

3.1 Methodology and Data Sources

In this section we quantitatively investigate the extent of VC investment in the U.S. LEO sector. As a baseline, we tracked VC activity in this sector from commercial private equity data provider Thomson ONE. Thomson ONE is widely used by academics and should provide a reliable dataset to study the sector. To capture deals that the database did not identify, we supplemented this material with Google searches.

We first identified all U.S. LEO deals in Thomson ONE through business description keyword searches. We obtained our list of keywords by reviewing the business descriptions of notable deals (e.g., Planet Labs, Skybox), as well as words or phrases that consistently appeared in our review of the literature. We then conducted a number of Google searches to identify any VC-backed LEO companies not captured in our Thomson ONE search.

With this list of preliminary "hits" from keyword searches, we next investigated company backgrounds to determine their relevance to LEO. After refining this list

---

112 We do note that while CrunchBase reports that Zero Gravity Solutions received $1.7 million in venture financing, it appears that such financing was not sourced from VC groups. See, Zero Gravity Solutions, Inc., SEC Filing Form D, July 16, 2014.

113 "ACME Advanced Materials, Inc. Announces First Commercial Production of 4" SiC Wafers in Microgravity." *PRWeb.com* (September 9, 2014).

of companies, we searched for investment specifics.[114] Key details included dates of investment, venture capital firms involved, and transaction amounts. For this information, we primarily used press releases and news articles, as well as Thomson ONE and another commercial database, Preqin, which also is a widely referenced source for private equity and venture capital activity. We also investigated the company's subsector within LEO.

It is important to emphasize that we focused exclusively on VC deals for this analysis. We excluded leveraged buyouts (LBOs), as well as deals that exclusively involved angel investors/angel networks or former venture capitalists and those that were solely financed through debt. In addition, we excluded grants and crowdsourced funding, such as those from Kickstarter, as well as funding from accelerator programs. Finally, we excluded deals where the transaction size was under $1 million. In each of these cases, the investment preferences would likely be unrepresentative of the VC community and the company's future viability more open to question.

While every effort was taken to identify all U.S. VC investments in LEO to date, certain investments may not have been identified given little public disclosure. Other startups may have been missed due to difficulty in identifying their involvement in LEO.

3.2 An Empirical Look at VC Activity in the U.S. LEO Sector

We break down our analysis into three categories: (A) number of transactions and amount invested over time, (B) LEO subsectors over time, and (C) identity of the investors. Broadly speaking, we have found an emergence of VC activity in the miniature satellite and launch subsectors since 2008. All other subsectors such as microgravity manufacturing and laboratory services, propulsion, and asteroid mining appear to compose roughly a quarter of deals and less than 10 percent of the equity invested.

Venture investments in LEO are dominated by Internet- and software-focused VC firms—as opposed to space "specialists,"—that often have partners with deep knowledge of the sector. While we discuss these industry trends at a more macro level in this section, transaction-level data and related details (investment dates, investors, amount invested, and subsectors) are summarized in **Appendix A** (pp. 110–112).

We stress that these figures (i.e., aggregate deal counts and amount invested) must be interpreted with caution given discrepancies among our data sources and the possibility that certain financing events were not captured in our analysis.

---

114 For all possible cases we verify investment details with press releases. In some cases, certain deals are contained in one database, while others are not. In these cases, we investigate whether such deals are merely identified differently or if they are unique. To do so, we search for company statements on the deal and ultimately use our discretion to determine if the deal fits our criteria for a VC investment.

CHAPTER 4 ■ Venture Capital Activity in the Low-Earth Orbit Sector

A. Number of Companies/Transactions and Amount Invested in LEO Sector (All Dates)

Number of Companies/Transactions

Our research suggests that 18 U.S. LEO-related companies have received VC financing to date, in a total of 37 transactions (including what appear to be three round extensions). Some companies were only involved in one transaction, others in several. As noted previously, we excluded leveraged buyout (LBO) deals from our dataset—for example, Thermo Capital Partners' acquisition of satellite communications company Globalstar in 2003 and Artemis Capital Partners' acquisition of sun sensor supplier Adcole Corporation in 2014. We also omitted investments solely by angels/angel networks and investments in non-U.S. LEO companies, such as UrtheCast, a Canadian company.

Figure 4.1 tracks VC transactions in the LEO sector and illustrates the growing interest in the sector. The red bars indicate the first identified transaction for the 18 U.S. companies, while the black bars include all 37 identified transactions (including three round extensions).[115] From 1983 to 1999, we identified only three companies received VC funding. Between 2000 and 2007, five firms were funded—more than in the prior 17 years. That number doubled to 10 from 2008 to just January 2015. Looking instead at all identified transactions (including round extensions) we find a similar pattern: four transactions from 1983 to 1999, nine transactions from 2000 to 2007, and 24 from 2008 to January 20, 2015.

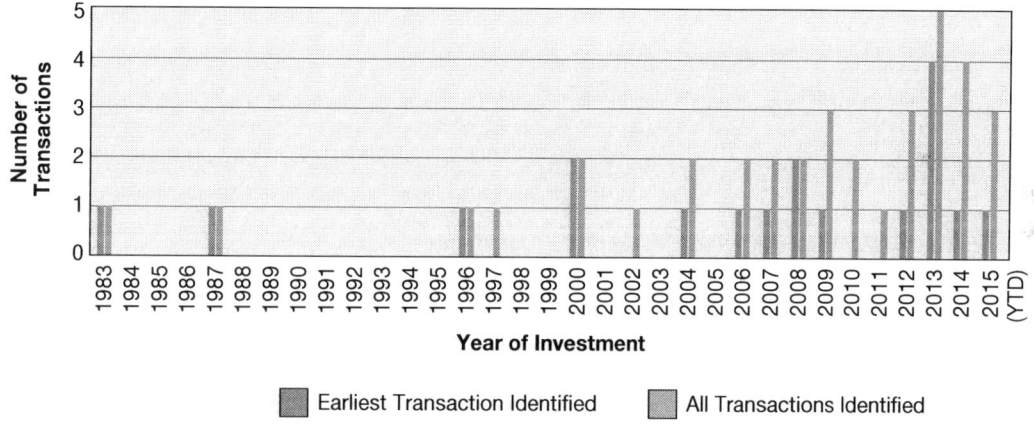

FIGURE 4.1: U.S. Venture Capital Transactions in LEO Sector, as at Jan. 20, 2015

---

115 We identify round extensions as investments that do not represent full funding rounds. If not explicitly stated, these typically are deals by current investors and represent insignificant amounts relative to the previous round.

Amount Invested

Among the subset of deals with transaction sizes disclosed (32 of 37), VC investment in the U.S. LEO sector totaled $1.64 billion (see **Figure 4.2**).[116] Given the small sample size, a few investments make up the majority of this figure (most notably, a $1 billion investment in SpaceX in 2015). Still, we find that in each five-year increment since 2000, aggregate transaction amounts have steadily increased: $89 million from 2000–2004; $184 million from 2005–2009; and $284 million from 2010–2014. In the first month of 2015, LEO investment levels skyrocketed as a result of Google and Fidelity's $1 billion investment in SpaceX. The SpaceX investment round accounts for over 60 percent of all VC funding to date in the U.S. LEO sector.[117]

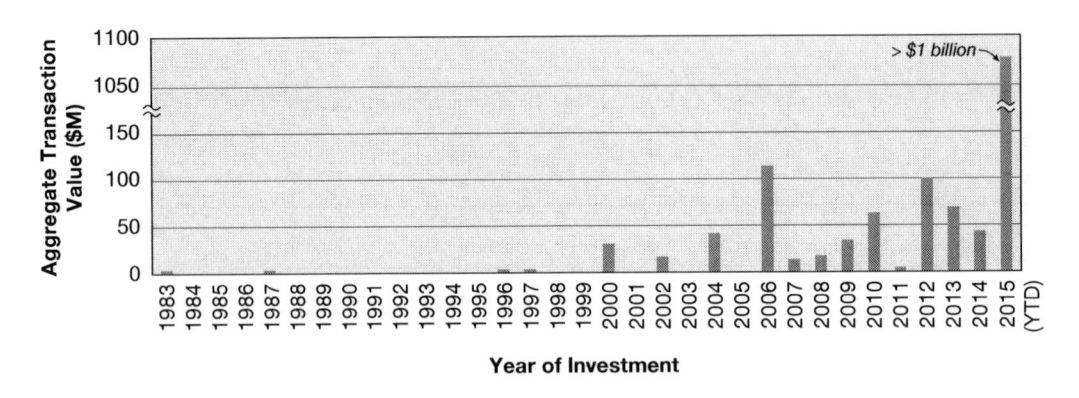

FIGURE 4.2: U.S. Venture Capital Transaction Value in LEO Sector, as at January 20, 2015

B. LEO Subsectors of Current Interest to VC Firms (2008–Jan. 20, 2015)

In this section, we take a closer look at the U.S. LEO companies in which VC firms have invested. The LEO sector can be broken down into a number of subsectors: satellites of varying sizes (e.g., nanosats, microsats) and for varying purposes (e.g., communication, Earth monitoring, space tourism); launch vehicles; commercial space services and scientific experimentation; propulsion; and asteroid mining. To see which of these subsectors has attracted the most attention from VCs, we break down deals and investment dollars by subsector since 2008.

Figure 4.3A charts the proportion of U.S. LEO VC deals by subsector. Because of a small sample size, we exclude deals that appear to be "round extensions," as they do not reflect the same type of VC interest as a full financing round.[118] We find that separate transactions in nanosatellites (Spire, Planet Labs) and microsatellites (Skybox)

---

116 Transaction values are missing in 2006, 2008, 2012, 2013, and 2014.

117 We include Google and Fidelity's $1 billion investment in SpaceX since appears to be structured as a standard VC deal and is reflective of rising VC interest.

118 We identify "round extensions" as investments that do not represent full funding rounds. If not explicitly stated in reports/databases as such, we define round extensions as deals by current investors that represent insignificant amounts relative to the previous round.

make up eight of the 21 deals, or 38 percent. We find another eight transactions in launch vehicles (SpaceX, Rocket Lab, Xcor Aerospace[119]), two in commercial laboratory services/microgravity-enabled technologies (NanoRacks; ACME Advancement Materials);[120] one in propulsion (Accion Systems); and two in asteroid mining (Planetary Resources).[121]

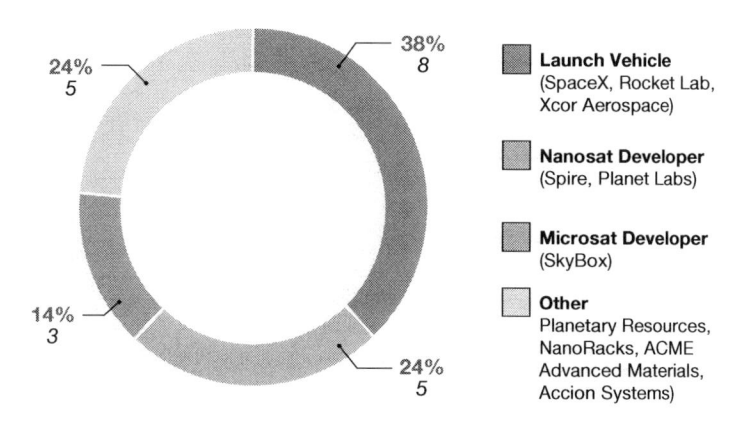

FIGURE 4.3A: Proportion of U.S. LEO VC Transactions (ex. Round Extensions), by LEO Subsector, 2008–January 20, 2015

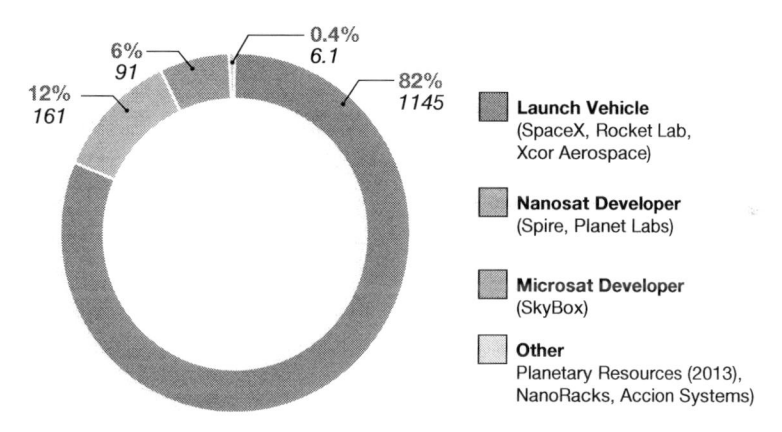

- 38% / 8 — Launch Vehicle (SpaceX, Rocket Lab, Xcor Aerospace)
- 24% / 5 — Nanosat Developer (Spire, Planet Labs)
- 14% / 3 — Microsat Developer (SkyBox)
- 24% / 5 — Other (Planetary Resources, NanoRacks, ACME Advanced Materials, Accion Systems)

FIGURE 4.3B: Proportion of U.S. LEO VC Investments ($M; Ex. Round Extensions), by LEO Subsector, 2008–January 20, 2015

- 82% / 1145 — Launch Vehicle (SpaceX, Rocket Lab, Xcor Aerospace)
- 0.4% / 6.1 — Nanosat Developer (Spire, Planet Labs)
- 6% / 91 — Microsat Developer (SkyBox)
- 12% / 161 — Other (Planetary Resources (2013), NanoRacks, Accion Systems)

---

119 We note that Xcor is also developing rocket propulsion systems.

120 After the date of this analysis, Spaceflight Industries, a launch service provider, announced a $20 million Series B funding round (March 2015) from RRE Venture Capital, Vulcan Capital, and Razor's Edge Ventures. Prior to this announcement, however, Series A funding ($7.5 million from Chugach Alaska Corporation and Apogee) appears to have been undisclosed.

121 Planetary Resources' financing history is not entirely transparent. We find it unclear as to whether the firm had two separate financing transactions in July 2013, as the company reportedly raised $1.5 million in a crowdfunding effort, as well as $1.5 million in "seed" funding during the same month from the investors noted in **Appendix A**.

In **Figure 4.3B** we break down VC activity by deal value. For consistency, we exclude round extensions, although they made up a trivial portion (less than 1 percent). It is important to note that Planetary Resources, Rocket Lab, Xcor Aerospace, and ACME Advancement Systems have not disclosed their respective VC funding amounts and thus are not included in **Figure 4.3B**. Among the 17 remaining deals, launch vehicles make up 82 percent of all deal value since 2008. We note, however, that this largely stems from SpaceX's $1 billion deal led by Google and Fidelity. Excluding this deal, nanosats and microsats represent 63 percent of investment dollars, launch vehicles account for 36 percent, and commercial lab facilities/microgravity-related products, propulsion, and asteroid mining under two percent combined.[122]

C. Key VC Firms Active in the LEO Sector

The VC backers of LEO companies have been dominated by six main groups: Bessemer Venture Partners, Draper Fisher Jurvetson (DFJ); Khosla Ventures; Founders Fund; E-Merge; and RRE Ventures. With the exception of Bessemer Venture Partners, each of these groups has reported at least two deals since 2008 in the LEO sector. While Bessemer Venture Partners (BVP) has only disclosed one deal to date, they are rumored to be involved in at least one other LEO company.[123]

Table 4.1 compares the industry preferences of these LEO-oriented VC firms. To do so, we tracked the proportion of deals since 2007 within Preqin's industry categories (which admittedly are slightly ambiguous). We find that these VC groups typically invest in Internet- and software-related companies. Among all deals from 2007 to present, the VC firms invested roughly 32 percent on average in "Internet" companies and 23 percent in "software and related" companies.[124] To put this into context, Preqin reported that among North American VC deals from 2008 to February 20, 2013, roughly 24 percent of deals were in "Internet" and 18 percent of deals were in "software and related" companies.[125] While the time periods do not align precisely, these figures suggest that the industry preferences of LEO-oriented VC firms may make them keenly alert to opportunities at the intersection in these sectors.

---

122 We again note that after the date of this analysis (March 2015), Spaceflight Industries, a launch service provider, announced a $20 million Series B funding round from RRE Venture Capital, Vulcan Capital, and Razor's Edge Ventures. Prior to this announcement, however, Series A funding ($7.5 million from Chugach Alaska Corporation and Apogee) appears to have been undisclosed.

123 Indeed, after date of this analysis BVP led Series B financing with Rocket Lab, which was announced on March 2, 2015.

124 These percentages represent a simple average since 2007 (i.e., average the proportion in these industries among the six groups). If we instead take a pooled average (i.e., dividing the total number of "Internet" and "software and related" deals among the six VC firms since 2007 by the total number of transactions during this period), we find that "Internet" companies composed 28 percent of transactions and software & related companies composed 18 percent of transactions.

125 Gemma Morris. "Venture Capital Deals: Industry Trends." In Private Equity Spotlight, March 2013: Preqin, 2013, Figure 4.

In Table 4.1 we note the partners from the firms who serve on the board of directors of their respective portfolio companies. Our research suggests that much interest among these firms in the LEO sector often originates from individual partners. One interviewee brought up the apparently personal nature of VC investing in LEO, noting, "It's not really that a firm has a practice in it. There is one partner who's interested in it." This toe-in-the-water approach is similar to the way in which many VC firms became involved in such then-unknown sectors as the nascent Internet in the early 1990s.

TABLE 4.1: Major VC Groups in U.S. LEO Sector[126]

NAME (YEAR FOUNDED)	LEO SECTOR INVESTMENTS	PARTNER SERVING (OR WHO SERVED) ON BOARD OF PORTFOLIO COMPANY	TOP THREE INDUSTRY PREFERENCES OF VC FIRM (2007–JAN. 19, 2015)
Bessemer Venture Partners (1911)[a]	Skybox	David Cowan (Skybox); Ethan Kurzweil (Board Observer on Skybox)	Total Deals: 428 Internet (27%) Software & Related (23%) Telecoms (11%)
Draper Fisher Jurvetson (1985)	SpaceX; Planet Labs	Steve Jurvetson (SpaceX; Planet Labs)	Total Deals: 548 Internet (27%) Telecoms (18%) Clean Tech (16%)
Khosla Ventures (2004)	Skybox; Rocket Lab	Pierre Lamond[b] (Skybox)	Total Deals: 382 Clean Tech (21%) Internet (16%) Software & Related (15%)
Founders Fund (2005)	SpaceX; Planet Labs; Accion Systems	Luke Nosek (SpaceX)	Total Deals: 122 Internet (48%) Software & Related (21%) Other IT (12%)
E-Merge (1998)	Spire; NanoRacks	N/A	Total Deals: 12 Software & Related (42%) Internet (33%) Industrials (17%)
RRE Ventures[c] (1994)	Spire; Accion Systems	N/A	Total Deals: 243 Internet (38%) Software & Related (23%) Telecoms (14%)

a After date of this analysis BVP led Series B financing with Rocket Lab, which was announced on March 2, 2015.
b Lamond formally left Khosla Ventures in June 2014.
c As previously noted, after the date of this analysis (March 2015), Spaceflight Industries, a launch service provider, announced a $20 million Series B funding round from RRE Venture Capital, Vulcan Capital, and Razor's Edge Ventures. Prior to this announcement, however, Series A funding from Chugach Alaska Corporation and Apogee appears to have been undisclosed.

---

126 Deal count data from Preqin, accessed January 19, 2015. Deal counts may not be exhaustive.

We find evidence of this phenomenon in our analysis: Khosla Venture's 2009 investment in nascent Skybox Imaging was primarily the work of then-Partner Pierre Lamond. One source commented that Lamond invested the first $3 million in Skybox Imaging after being charged by Vinod Khosla, founder of Khosla Ventures, to identify high-risk science ventures in which the firm could invest a quarter of its $1 billion fund.[127] In addition, Steve Jurvetson—a long-time rocket enthusiast—represents Draper Fisher Jurvetson (DFJ) on the board of its two LEO sector investments. In fact, at the 2014 "Small Satellite Conference," Jurvetson explained that out of "personal curiosity" he had been meeting with entrepreneurs in the space-related industries for 10 years before DFJ's investment in SpaceX.[128]

We do find, however, recent trends suggesting that the space sector is becoming more a firm-wide area of focus. In one such example, Bessemer Venture Partners, backers of SkyBox and, recently, Rocket Lab, has announced a "spacetech" practice that includes David Cowan, Sunil Nagaraj, and Scott Smith, the COO and board member of Iridium Communications.[129]

Section 4. Performance of Venture Capital Investment in LEO

4.1 Setting the Context: The Importance of Strong Management Teams and the "Early Performers" in Young Industries

In this section, we look at the performance of VC-backed firms in the LEO sector. Before we explain our methodology, however, we emphasize that early success in the industry would be a testament to venture capitalists' abilities to partner with high potential, commercially driven firms, not merely fascinating technologies. As noted in **Section 1.2**, VC firms conduct extensive due diligence to assess management team strength and market attractiveness.

A large amount of empirical literature further links managerial strength and VC financing. A classic study by Tyzoon Tyebjee and Albert Bruno published in 1984 found that managerial capabilities had the strongest effect among all evaluation

---

127 David Samuels. "Inside a Startup's Plan to Turn a Swarm of DIY Satellites into an all-Seeing Eye." *Wired.com* (June 18, 2013).

128 See "Small Sat 2014: Keynote Steve Jurvetson." YouTube video, 1:08:27. Posted by Small Sat Conference, August 18, 2014. *https://www.youtube.com/watch?v=qzudBqGyPTY#t=340* [5:40–8:00].

129 "Iridium COO Scott Smith Joins Bessemer Venture Partners as an Operating Partner." *BusinessWire* (March 2, 2015).

dimensions on reducing the riskiness of a deal.[130] In addition, a number of more recent studies help substantiate this finding:

- Steven Kaplan and Per Strömberg studied 67 portfolio investments made by 11 VC firms and found that managerial and leadership experience is a crucial part of the VC decision-making process. In fact, venture capitalists cited management quality as a reason for investing in nearly 60 percent of the investments.[131]
- Judith Behrens, et al. used data from 138 VC financing rounds in the United States and Europe and found that young biopharmaceutical ventures run by management teams with educational backgrounds in the fields of management, law, medicine, and biosciences acquire more money in VC financing rounds.[132]
- Eli Gimmon and Jonathan Levie studied 193 high-tech startups that were participants in the Israeli Technology Incubator Program and found that new high-tech ventures whose founders have business management expertise had better odds of attracting venture funding.[133]

Venture investors also seek companies operating in large, expanding markets. Tyebjee and Bruno's study (referenced above) found that market attractiveness (size, growth, and access to customers) was the strongest driver of expected returns for venture capitalists. Kaplan and Strömberg's study (also referenced above) reported that market size was cited as an important factor in close to 70 percent of investments in the sample. As a result, high-performing "role-model" ventures are especially critical to the development of nascent markets. This phenomenon is noted by entrepreneurship professor Lowell Busenitz in his examination of venture capital's ability to develop new industries: "When there are no industry benchmarks, no established ventures in a specific industry and there is no clear market, most venture capitalists seem to be very reluctant to pursue such opportunities."[134]

---

130 The authors studied five major "dimensions" (market attractiveness, product differentiation, managerial capabilities, environmental threat resistance, cash-out potential) of expected return and perceived risk. See Tyzoon Tyebjee and Albert Bruno. "A Model of Venture Capital Investment Activity." *Management Science* 30, no. 6 (1984): 1051–1066, 1060.

131 Interestingly, the study also found that this education could have a negative effect on VC funding for older ventures. Steven N. Kaplan and Per Strömberg. "Characteristics, Contracts, and Actions: Evidence from Venture Capitalist Analyses." *The Journal of Finance* 59, no. 5 (2004): 2177–2210.

132 Judith Behrens, Holger Patzelt, Lars Schweizer, and Robin Bürger. "Specific Managerial Human Capital, Firm Age, and Venture Capital Financing of Biopharmaceutical Ventures: A Contingency Approach." *Journal of High Technology Management Research* 23, no. 2 (2012): 112–121.

133 Eli Gimmon and Jonathan Levie. "Founder's Human Capital, External Investment, and the Survival of New High-Technology Ventures." *Research Policy* 39, no. 9 (2010): 1214–1226, 1222.

134 Lowell W. Busenitz. "Innovation and Performance Implications of Venture Capital Involvement in the Ventures they Fund." In *Handbook of Research on Venture Capital*, edited by Hans Landström, 219–235. Northampton, MA: Edward Elgar Publishing, 2007.

In other words, infant markets without any so-called "shining stars" or notable exits are largely unattractive to venture capitalists. Irrespective of the innovative capabilities of a given technology or the visions of a strong management team, venture investors only back companies that they believe to be operating in a perceived healthy market. As a result, a major exit representing a theretofore-untapped opportunity would likely have a mushrooming effect in the VC community.

4.2 Methodology

With the importance of early performance for the LEO sector in mind, we need to determine the prospects of startups currently operating in the sector. While an in-depth analysis of the stock of entrepreneurial talent in the LEO market is outside the scope of the study, we seek to identify the performance to date of emerging VC-backed companies. Because the private LEO market was largely undefined prior to the funding of SpaceX—with a few notable examples indicated in **Appendix A**—we suggest that successes ("role models") in our small sample of VC-backed U.S. startups could draw more mainstream involvement in the sector.

Determining the performance of private companies is notoriously difficult. Ideally, we examine the exit results of these investments and the gains that accrued to the investors when the companies were acquired or went public. Many of these companies, though, are very young and have not had either an exit event or a financing round at such an impressive figure that additional information was forthcoming. The valuation of companies that raise subsequent financing rounds (B, C, and so forth) can suggest success or failure, as can whether the transaction attracts new backers. But this information is not always easy to find. VC firms often do their best to obscure the valuation at which their companies are funded, unless doing so is to their advantage. Thus, we are constrained by the availability of the data.

For this assessment, we turn again to publicly available data from press releases, news articles, Thomson ONE, and Preqin. We note again that we exclude all non-VC investments (i.e., LBOs, angel investors/former venture capitalists, corporations, and so on) from this analysis. We include deals as of 2008 (i.e., SpaceX's first external investment), as these companies best reflect the current direction of the industry. We exclude deals that occurred after 2014, because there would be no follow-on funding data as of the date of our writing. We then measure performance in two ways, which we describe below.

A. Successive Funding Rounds

As discussed in **Section 1.2**, venture capitalists typically structure deals to reward the achievement of milestones. This "staged" model of financing helps venture capitalists control for costly, private information held by entrepreneurs (information asymmetries), as well as the high levels of uncertainty associated with the value of intangible assets—often intellectual property or trade secrets—held by startups.[135] Using this approach, VC firms can optimize investment strategies as technologies and markets are proven.[136]

It is important to note, however, that more funding rounds do not necessarily imply greater success, as research suggests that ventures that succeed quickly would likely receive fewer rounds of funding. In other words, startups with low probabilities of failure would likely lead venture investors to avoid the costs associated with renewal of funding, such as contract negotiations, lawyer fees, and time spent on evaluations[137] and instead provide the startup with more capital within each round.[138] Given this insight, we only consider whether the company achieved multiple rounds of financing—i.e., not the number of investment rounds.

Second, we look at the size of successive investments. Research suggests that thriving startups tend to receive larger investments in each round, as the value of abandonment diminishes for the VC firm.[139]

B. Exit Via Acquisition or IPO

The performance of a venture-backed company is also indicated by a successful exit. Venture capitalists only invest in companies with the belief that their equity stakes can be liquidated at a premium. Josh Lerner, et al. explain, "Companies accrete value in specific steps [in such a way that they]...reach the point where the gains due to

---

135 Paul A. Gompers. "Optimal Investment, Monitoring, and the Staging of Venture Capital." *The Journal of Finance* 50, no. 5 (Dec., 1995): 1461–1489. With regard to asset intangibility, Gompers writes, "As asset tangibility and liquidation value increase, venture capitalists can recover more of their money if liquidation occurs, and the need to monitor declines. By gathering information, venture capitalists determine whether projects are likely to succeed and continue funding only those that have high potential" (p. 1484).

136 This idea is discussed in, Dirk Bergemann, Ulrich Hege, and Liang Peng. "Venture Capital and Sequential Investments." Cowles Foundation Discussion Paper no. 1682R (March 2009).

137 Gompers discusses these transaction costs in more detail in, Paul A. Gompers. "Optimal Investment, Monitoring, and the Staging of Venture Capital." *The Journal of Finance* 50, no. 5 (December 1995): 1464.

138 This insight is drawn from Dirk Bergemann, Ulrich Hege, and Liang Peng. "Venture Capital and Sequential Investments." Cowles Foundation Discussion Paper no. 1682R (March, 2009). In their theoretical analysis, the authors explain, "As the project advances and the probability of eventual success increases, investment flows should be optimally increasing.... The optimal staging sequence depends on the value of the real option to abandon: The higher the estimated final value of the project, and the larger the estimated success probability, the fewer rounds will be used.... At the same time...the capital allocation for each of these rounds will increase" (p. 35). They support these findings in their empirical analysis of U.S. VC investments from 1987 to 2002.

139 Dirk Bergemann, Ulrich Hege, and Liang Peng. "Venture Capital and Sequential Investments." Cowles Foundation Discussion Paper no. 1682R (March 2009).

additional time under the current investors flatten out."[140] Exit routes can take a variety of forms: IPO, acquisition, secondary sale, or shutdown.

In line with recent academic research from Shai Bernstein, et al., we identify "successful" exits as exits via IPO or acquisition.[141] It is important to note that because the average time to exit for the U.S. VC industry in aggregate is roughly 5.9 years among M&A exits and 7.2 years among IPO exits from 2010 to 2013, we expect to see few exits for these companies.[142] We also search for any bankruptcies to date.

While the outcomes of many of the privately held firms remains uncertain, our analysis of the results to date should be reflective of the ultimate performance.

4.3   Performance of Recent U.S. LEO Companies (2008–2014)

Table 4.2 summarizes the performance of recent LEO deals (excluding round extensions). Because of the unclear nature of its funding and contradictory media reports, we exclude Planetary Resources from this analysis. We find that of the remaining eight companies that received VC funding between 2008 and 2014, five (SpaceX, Skybox Imaging, Spire, Planet Labs, and Xcor Aerospace) have raised additional VC funding as of Jan. 20, 2015.[143] We also note that all three deals without follow-on funding took place in 2013 or 2014, giving them a lower probability of second rounds within the sample. We find that the size of the transaction generally increased in sequential rounds.

Furthermore, the companies that have yet to receive an additional round of VC funding have either received other types of funding or achieved significant milestones that validate their technology:

- NanoRacks helped validate its technology by successfully deploying Planet Labs' Flock 1 CubeSats from the ISS in February 2014.[144]

- Rocket Lab has received grants from the New Zealand government in January 2014 and completed an important developmental step for its carbon composite Electron rocket, in September 2014.[145]

---

140   Josh Lerner, Ann Leamon, and Felda Hardymon. *Venture Capital, Private Equity, and the Financing of Entrepreneurship*. New York: John Wiley & Sons, Inc., 2012, p. 200.

141   Shai Bernstein, Xavier Giroud, and Richard Townsend. "The Impact of Venture Capital Monitoring: Evidence from a Natural Experiment." (February 23, 2014).

142   Thomson Reuters. *National Venture Capital Association Yearbook 2014*. Arlington, NVCA, 2014, p. 73, 77.

143   We note that Xcor Aerospace's second deal was not from a traditional VC group. See **Appendix A** for more details.

144   Irene Klotz. "Satellite 'Flock' Launched from ISS Cubesat Cannon: Photos." *News.Discovery.com* (February 18, 2014).

145   "NZ Govt. Funding Secured." *RocklabUSA.com* (January 2014). See also, "First Off-Tool Fairing Complete." *RocklabUSA.com* (September 2014).

- ACME Advancement Materials (A2M), which "...develops and produces unique materials in a microgravity environment," announced the ability to produce its Silicon Carbide wafers in "commercially viable quantities" in October 2014.

We also find one exit via strategic acquisition, which suggests viability in the sector for venture capitalists. We emphasize that Google's acquisition of SkyBox demonstrates large-scale commercial applications of the type of data small satellites can acquire and will likely serve as a benchmark against which other satellite companies can be assessed.

TABLE 4.2: Amount Invested ($ Million) by Round of U.S. LEO VC Companies (Ex. Round Extensions), 2008–January 20, 2015

COMPANY NAME	ROUND 1	ROUND 2	ROUND 3	ROUND 4	ROUND 5	EXIT
Xcor Aerospace[a]	Aug. 2008 Equity Inv.: N/A	— Note: Non-VC funding (Feb. 2012)	May 2014 Equity Inv.: 14.2	—	—	—
SpaceX	Aug. 2008 Equity Inv.: 20.0	Aug. 2009 Equity Inv.: 30.4	Nov. 2010 Equity Inv.: 50.2	Dec. 2012 Equity Inv.: 30.0	Jan. 2015 Equity Inv.: 1,000.0	—
Skybox Imaging	July 2009 Equity Inv.: 3.0	July 2010 Equity Inv.: 18.0	April 2012 Equity Inv. 70.0	—	—	Aug. 2014 Acquisition by Google ($500M)
Spire (formerly Nanosatisfi)	Feb. 2013 Equity Inv.: 1.2	July 2014 Equity Inv.: 25.0	—	—	—	—
Planet Labs (formerly Cosmogia)	June 2013 Equity Inv.: 13.1	Dec. 2013 Equity Inv.: 52.0	Jan. 2015 Equity Inv.: 70.0	—	—	—
NanoRacks	June 2013 Equity Inv.: 2.6	—	—	—	—	—
Rocket Lab[b]	Oct. 2013 Equity Inv.: N/A	—	—	—	—	—
ACME Advancement Materials	Feb. 2014 Equity Inv.: N/A (note: reported as "seven-figure")[c]	—	—	—	—	—

a   For more details on Xcor Aerospace's VC funding, see Appendix A.
b   As previously noted, Rocket Lab indeed received Series B financing in a round led by BVP announced on March 2, 2015.
c   Kevin Robinson-Avila. "Made in Space." *Albuquerque Journal* (October 13, 2014).

Section 5. The Logistics of Future Involvement

In this section we touch on a few points to emphasize the issues currently inhibiting venture capital development in LEO, ISS-connected companies. We follow with a set of recommendations on how NASA/CASIS can best leverage the ISS to boost VC activity in LEO.

5.1 The Overarching Issues

In its efforts to stimulate VC investment in LEO, NASA faces several interlinked challenges. It is clear that venture capitalists are interested in LEO, as seen by over $250 million of equity invested in miniature satellites and over $1 billion in launch vehicles since 2008. The emergent sector has attracted widespread media attention, with Google's $500 million acquisition of Skybox and SpaceX's $12 billion valuation. Entrepreneurial development in ancillary services, such as launch vehicles and propulsion systems dedicated to miniature satellites, as well as favorable regulatory reforms, suggest continued innovation in this arena.

We find, however, that there is much less VC interest directed specifically at the ISS for several reasons. These include a lack of demonstrated commercial success in entrepreneurial microgravity manufacturing; limited opportunities to access the ISS; the physical constraints of the ISS; and lack of awareness of ISS as a viable national lab for applied commercial research.

We describe these and some possible responses below and follow with recommendations for future action.

A. Lack of Demonstrated Commercial Success in Entrepreneurial Microgravity Manufacturing

Venture capitalists very rarely invest in technology qua technology. They invest in companies, which include technology but house it within the context of a management team and a defined market. Within the VC community, there is a lively debate regarding whether one backs the horse (technology), the jockey (the management team), or the race course (the market). Venture capitalists have backed world-changing companies by pursuing any of these three strategies, but in general, the management team is viewed as among the top, if not the most important, component of a startup. A talented team, it is argued, can respond to failure in the technology or changes in the market, while a weak team may trip up a good technology or fail to address obvious market needs. Microsoft's IP Ventures effort, for example, serves as a fitting case study: Microsoft sought to spin out innovative technology that did not fit its product road maps in exchange for an equity share in the operation, but found no interest until it also supplied a management team.[146]

Thus, for example, although research shows how a microgravity environment may be conducive to enhanced protein crystallization and cell and tissue culturing, each of which has clear commercial applications in biotechnology, venture capitalists will not be interested in investing in such research in and of itself. The process would need to have a start-up company pursuing a commercial application and a strong likelihood of creating something—whether a product or a method—that could be sold. Historically, those VC firms that focused on applied materials soon moved into more commercial applications, as one GP noted: "You may have a great chemical compound but the money is when the compound is put into paint or roofing or solar collectors that people can buy."[147]

B. Limited Opportunities to Access the ISS

The complexities inherent in accessing space manufacturing inhibit entrepreneurial activity in this area. While NanoRacks coordinates spaceflights on virtually any cargo ship to the space station and develops innovative platforms (most notably, its NanoLab payload hardware permanently installed on the U.S. National Laboratory) on which entrepreneurial firms can conduct microgravity research, official documentation suggests it takes between nine and 14 months on average from contract signing to launch for its research laboratory services (taking into account the time needed to comply with NASA regulations).[148]

As alluded to by many of our interviewees, a robust emerging entrepreneurial market for LEO R&D would require much more readily available access. Currently, NanoRacks supplies secondary payload opportunities to the ISS via all six launch vehicles—but only one, the SpaceX Dragon, can presently return substantial cargo back to earth (Soyuz has limited capacity and others burn up in the atmosphere). Such cargo, though, may be absolutely critical for further analysis—or for supplying the product that is to be sold.[149] In fact, of the roughly 740 kg of research resources brought to the ISS ("upmass") from September 2013 to March 2014, only 38 kg was returned ("downmass").[150] We do note, however, that CASIS is working with the

---

146 Josh Lerner and Ann Leamon. "Microsoft's IP Ventures" Harvard Business School Case no. 810-096 (Boston: Harvard Business School Publishing, 2011).

147 Personal conversation with GP at Ampersand.

148 For details, see *http://nanoracks.com/wp-content/uploads/NanoRacks-Commercial-Spacelab-Presentation-1.pdf*. See also, NanoRacks. NanoRacks CubeSat Deployer (NRCSD) Interface Control Document. Houston, NanoRacks, 2013.

149 National Aeronautics and Space Administration. NASA's Efforts to Maximize Research on the International Space Station: NASA Office of Inspector General, 2013, p. III. See also, Steven Clark. "Commercial Dragon Supply Ship Returns to Earth." *Spaceflightnow.com* (February 1, 2015).

150 International Space Station Utilization Statistics Expeditions 0–38: December 1998–March 2014. NASA, August 2014.

private sector to enable more rapid sample return and NASA has awarded the Sierra Nevada Corporation a Commercial Resupply Services 2 contract for the services of its Dreamchaser vehicle with payload return capabilities.[151]

This need for more frequent access to LEO is attracting substantial attention on the part of the miniature satellite market. As emphasized in **Table 4.1**, dedicated launch vehicles such as VC-backed Rocket Lab are being developed to provide weekly LEO access. According to Rocket Lab CEO Peter Beck, "We can't do anything substantial if we can't get the frequency up.... There are all sorts of inspiring things people want to do in space, but it's hard if we can't increase launch capacity and reduce cost."[152] In fact, one of the venture capitalists in our interviews explained that space access has been the single largest inhibitor to space entrepreneurship. As a result, we find that along with other issues to be discussed below, the mere ability to access the ISS is a substantial obstacle to attracting entrepreneurial and venture capital interest in the ISS.

C. Physical and Time Constraints on ISS

The ISS is also replete with physical limitations. Port capacity is limited, so additional launches will eventually be unable to offload their cargo quickly. The orbit is at an angle and a speed that presents challenges to commercial Earth observation. Crew time is already constrained, limiting the number of additional experiments that can be undertaken. As noted by an ISS research scientist, while the station's total research capacity is underutilized in terms of occupancy of allocated space, personnel is limited: "We're oversubscribed in crew time: we actually have more things people would like to do than the crew has time to help with."[153] More specifically, roughly 81 percent of the internal space allocated for research in the U.S. portion of the ISS was occupied from April 2013 to March 2014. Moreover, roughly 37 percent of external sites—which, according to NASA, are generally used for "astronomical studies, Earth observation and technology development and demonstration for robotics, materials, and space systems"—were occupied during the same period. While NASA aims to allocate about 35 hours per week of crew time to research-related activities, this figure was over 40 on average from April 2013 to March 2014.[154] Despite this effort, however, crewmembers may lack the specialized training to ensure that the experiments yield the most critical results.

All of these constraints could be solved—additional ports, additional dormitory space—but all require additional investment. Here ISS encounters another constraint: its time horizon. The United States has committed to ISS operations through 2024,

---

151 Jeff Foust. "Commercial Vehicle Promises More Frequent Return of ISS Experiments." *Spacenews.com* (October 31, 2014).

152 See Khosla Venture's Rocket Lab cast study, at *http://www.khoslaventures.com/portfolio/rocket-lab*.

153 Jeff Foust. "Making the most of the ISS." *Thespacereview.com* (March 24, 2014).

154 National Aeronautics and Space Administration. *Extending the Operational Life of the International Space Station Until 2024*. Washington, DC: NASA Office of Inspector, 2014.

and operations are likely to be technically feasible for a number of years after that, but there is recognition that at some point in the future there is most likely going to be an end-of-life (EOL) determination for the ISS, at least in its current configuration.

D. Lack Of Awareness of ISS as a Viable National Lab for Applied Commercial Research

For much of the VC community, ISS is not recognized as a national lab along the lines of Lawrence Livermore or Los Alamos, which are themselves rarely considered as possible locations for civilian entrepreneurs to do research. Research tends to spin out of these national labs into entrepreneurial ventures; not the other way around. As noted in NASA Office of the Inspector General's Audit Report, "… the majority of the research activities conducted aboard the ISS have related to basic research as opposed to applied research."[155] Even NanoRacks, which created commercial research modules inside the ISS, has found miniature satellite deployment to be its most popular service.[156]

5.2 Our Recommendations to Spur Economic Development Via ISS

In light of our analysis we offer five recommendations of future action to bring together venture capitalists and LEO entrepreneurs. Our main recommendations focus on three types of efforts: awareness, knowledge, and funding. More specifically, we suggest:

- **Awareness:** Continued efforts to raise awareness among entrepreneurs and venture capitalists regarding the intersections of the ISS with such industries as biotechnology and miniature satellites, as well as the steps needed to use the ISS efficiently.

- **Knowledge:** A venture capital advisory committee offering feedback to senior management at NASA on how entrepreneurs could utilize the ISS.

---

[155] National Aeronautics and Space Administration. *Extending the Operational Life of the International Space Station Until 2024*. Washington, DC: NASA Office of Inspector, 2014, p. 47.

[156] Jeff Foust. "Space Station's Commercial Users Hitting Bottlenecks." *Spacenews.com* (February 27, 2015).

- **Investment:** Investment could come in two forms. First, continued collaboration with relevant angel networks. Second, NASA could consider its own accelerator program geared towards ISS-connected companies.[157]

We rank our suggestions for addressing these concerns from shorter-term to longer-term and explain each below. It is important to note that these recommendations are not mutually exclusive.

A. Raise Awareness Among LEO-Oriented Entrepreneurs and Relevant Angel Networks/VC Groups of ISS and Its Benefits, and How to Make Use of It

In line with CASIS' overarching mission, we suggest continued effort to promote awareness among entrepreneurs of the ISS as a national laboratory and the role it can play to speed research for entrepreneurial efforts. The recent partnership between CASIS and the Space Commerce Conference and Exposition (SpaceCom), as well as its sponsorship of the ISS R&D Conference represent the type of awareness campaigns that could catalyze private sector involvement in space.

We emphasize, however, the importance of continued attendance at professional gatherings outside of those specifically dedicated to space commerce. We support CASIS' current involvement in such trade shows as the BIO International Convention and the World Stem Cell Summit, and suggest involvement in similar opportunities dedicated to each area of applied science that could benefit from a microgravity environment. We also note the importance of publicizing projects that have succeeded more quickly due to research done in microgravity.

Few venture capitalists recognize the extent to which the ISS could serve as a platform for applied research breakthroughs for startups. We therefore also suggest involvement in major angel and VC conferences to spark interest in this community. Critically, such involvement need not—in fact, should not—be solely dedicated to space-related groups, but instead also to groups specializing in such areas as life sciences and physical sciences, as well as the software-related groups that have already demonstrated an interest in miniature satellite companies and the vast amounts of data that they generate.

As an area of future research, we also suggest a survey of startups in industries that could potentially benefit from a microgravity environment to quantify the level of awareness of the ISS as a national research platform. Perhaps certain industries are keenly aware of CASIS and the steps necessary to access the ISS, while others are less so. A diagnostic test—for example, a simple three question survey handed out during

---

157 An accelerator provides space, mentorship, and a small amount of money, which may be a loan, grant, or equity investment, to very young startups. Angel investors typically are individuals or small groups who have not raised a formal fund. They generally invest in the form of a subordinated loan that converts to equity at the terms of the first formal VC investment. A VC investment tends to be standard investment where the investor provides cash and advice in exchange for an equity position in the company. Not all three of these funding strategies occur in a given start-up: some may go directly to VC investment; others may take VC investment while in an accelerator; yet others receive angel funding and never raise subsequent money.

conferences—measuring awareness levels among these groups may help optimize a marketing strategy. We could not find any such data as of the date of our writing.

B. Set Up a Committee of Venture Capitalists to Advise NASA Management on Strategy of Private Sector ISS Involvement

We also suggest that NASA and/or CASIS establish a VC advisory committee to guide the LEO commercialization strategy. Venture capital advisors could (i) identify current trends in LEO entrepreneurship; (ii) provide subtle knowledge of the challenges that entrepreneurs encounter with the ISS and advise management on how to optimally address these; and (iii) raise the VC industry's awareness of the ISS, CASIS, and NASA by providing a contact point. We examine points i and ii in more detail below.

i. Identification of Current Trends in LEO Entrepreneurship

Venture capitalists are aware of the most recent entrepreneurial trends. First, venture capitalists examine hundreds or thousands of startup solicitations each year.[158] As a result, they possess knowledge of the emerging types of LEO-oriented companies, whether they be miniature satellites or microgravity-enabled pharmaceuticals.[159]

Gompers, et al. lend empirical support to the agile market responsiveness of "specialist" venture capitalists—that is, those that specialize in certain industries (software, hardware, life sciences, and so on). The authors found that in light of positive market signals about the attractiveness of certain sectors, "specialist" venture capitalists respond favorably by increasing their investment rates.[160]

Armed with this knowledge, CASIS management could orient ISS marketing/awareness efforts to these subsectors, as well as align initiatives on the ISS to intersect with entrepreneurial interest.

ii. Subtle Knowledge of the Challenges that Entrepreneurs Encounter with the ISS and Advice on How to Address Them

Venture capitalists are also in a strong position to advise NASA and/or CASIS management on the specific challenges that entrepreneurs face in the LEO sector. They acquire this knowledge in the due diligence process, as well as in their active involvement with portfolio companies.

---

158 David Kirsch, Brent Goldfarb, and Azi Gera. "Form Or Substance: The Role of Business Plans in Venture Capital Decision Making." *Strategy Management Journal* 30, no. 5 (2009): 487–515.

159 According to William Meehan III, et al., "VC investment activity provides outsiders with early signs of key trends emerging in high tech, communications, and biotechnology. These early signs can help executives monitor technology transitions that may fundamentally disrupt business processes…." See, William F. Meehan III, Ron Lemmens, and Matthew R. Cohler. "What Venture Trends can Tell You." *HBR.org* (July 2003).

160 Paul Gompers, Anna Kovner, and Josh Lerner. "Specialization and Success: Evidence from Venture Capital." *Journal of Economics & Management Strategy* 18, no. 3 (2009): 817–844.

For each startup that passes the "screening" and "pitch" phases, venture capitalists invest significant time to study (among many other details) whether the company operates in a growing market, has a solid base of potential customers, and possesses adequate distribution channels to reach customers. For the roughly one percent of companies in which the VC firm invests, venture investors serve on boards to professionalize operations, refine strategy, and consider exit options.

NASA/CASIS could leverage such knowledge to identify challenges that LEO entrepreneurs commonly face and find ways of addressing these issues from the viewpoint of the financers of these entrepreneurs. Such information may shape awareness strategies, spawn new initiatives, and/or retire ineffective projects.

It is critical to recognize that the best VC firms are those least likely to be able to dedicate substantial time to a project like this. Should this recommendation be pursued, we highly recommend that meetings be held twice a year for no more than an hour at a time. Meetings should be guided with a detailed agenda and it should be expected that many of the venture capitalists will attend via telephone.

C. Expand Collaboration with Angel Networks

Another important initiative involves partnering with and funding relevant angel networks. Angel networks match entrepreneurs with angel investors who can provide crucial seed/early stage funding for startups. We understand that CASIS has partnered with the Space Angels Network and the Houston Angel Network to provide introductions to companies with promising ideas. We applaud these sponsorships and see the potential for others.

We emphasize that academic research supports the complementary relationship between angel investment and VC investment. As explained by Andrew Wong, et al. who examined a dataset of angel-backed firms between 1994 and 2001, "Angels take on more risks and invest smaller amounts in younger firms than venture capitalists. Angel investors appear to nurture younger firms until the company is established enough for venture consideration."[161] This complementary relationship has also been found in the UK context, as, for example, survey results in one study suggested that angel-backing in itself was often perceived as a positive signal by VC firms (45%) and that cross-referring investment opportunities was common.[162]

Because many LEO-related companies are often compete with terrestrial peers (e.g., analytics, vaccines, pharmaceuticals), we suggest that CASIS increase involvement with angel networks that can provide expertise in these domains. For example,

---

161 Andrew Wong, Mihir Bhatia, and Zachary Freeman. "Angel Finance: The Other Venture Capital." *Strategic Change* 18, no. 7 (2009): 221–230.

162 We do note that 5 percent of respondents generated perceptions depending on the business angel and the remaining half stated angel backing had no bearing on an investment decision. Richard T. Harrison and Colin M. Mason. "Venture Capital Market Complementarities: The Links between Business Angels and Venture Capital Funds in the United Kingdom." *Venture Capital* 2, no. 3 (2000): 223–242.

software and "big data" expertise is critical to miniature satellite companies (such as Skybox, Planet Labs, Spire), while semiconductor expertise is critical to companies developing wafers in microgravity (such as A2M). Given "space-side" and "terrestrial-application" angel expertise, angel-funded LEO startups will more likely attract VC and be better equipped to commercialize.

D. Catalyze Private Sector Involvement Via Accelerator Program Partnerships for ISS-Oriented Firms

Our final recommendation would be to continue to fund seed stage commercial ventures via partnerships with seed accelerator programs (like the MassChallenge Startup Accelerator).[163] As it stands, angel investors and independent venture capitalists back companies, not technologies. This idea was consistent with our interviews and is explicitly noted by CASIS' angel network partners. The Space Angels Network notes, "… members generally will not fund ideas or technologies."[164] Similarly, Houston Angel Network notes, "In order to be seriously considered, a company must have a completed, working prototype and have market validation (pilot, beta users, revenue)."[165] Accelerators may offer to early, "formation-stage" startups the resources to reach this stage of development.

While academic literature has been dedicated to the learning effects of accelerators, we find the most telling way to evaluate such programs is the level of follow-on funding raised by the accelerated ventures.[166] As explained by Fehder and Hochberg, VC investors view accelerators as "deal sorters" and "deal aggregators"; in other words, accelerators screen and co-locate a large population of startups to reduce search costs for venture capitalists.[167] An emerging strand of academic literature in

---

163 By way of background, we note that a seed accelerator is formally defined as "[a] fixed-term, cohort-based program, including mentorship and educational components, that culminates in a public pitch event or demo-day." It is important to clearly identify the differences between angel investments and accelerator programs. Most generally, whereas angel investors generally provide longer term investments with minimal education and mentorship given on an as-needed basis, accelerator programs are short-duration programs (often three months) that offer extensive mentorship (often with venture capitalists) and typically culminate in "demo days" where founders pitch businesses to potential investors. See, Susan G. Cohen and Yael V. Hochberg. "Accelerating Startups: The Seed Accelerator Phenomenon." Working Paper (March, 2014).

164 See Web site, at *http://spaceangelsnetwork.com/selection-criteria*.

165 See Web site, at *http://houstonanglenetwork.weebly.com/entrepreneurs.html*.

166 An example of such literature is Susan L. Cohen and Christopher B. Bingham. "How to Accelerate Learning: Entrepreneurial Ventures Participating in Accelerator Programs." *Academy of Management Proceedings* (January, 2013).

167 Daniel C. Fehder and Yael V. Hochberg. "Accelerators and the Regional Supply of Venture Capital Investment." Working Paper (September 19, 2014).

fact has found that accelerator-backed startups achieve higher levels of VC financing than their non-accelerator-backed counterparts.[168]

Because accelerators may be the missing link in the chain between idea generation and angel and/or VC investment, we suggest expanded collaboration with a geographically diverse base of accelerator programs. By partnering with accelerators willing to fund projects ranging from life sciences to earth observation/remote sensing and offering additional funding to ship technologies to the ISS, we suggest that promising LEO ventures will have the external validation and proofs of concept necessary to entice angel investors and venture capitalists. More specifically, we suggest that accelerated projects should receive preferential access to the ISS as well as highly subsidized (50–100 percent) launches. From the perspective of the entrepreneurs, these accommodations would knock down several barriers inhibiting VC financing. At the same time, CASIS would be investing in the companies most likely to achieve commercialization.

Section 6. Conclusion

In its efforts to increase the commercial use of the ISS, NASA faces several challenges. Keen venture capital interest exists in commercial opportunities in LEO, as shown in SpaceX's $12 billion valuation, Skybox's $500 million acquisition price tag, and the other companies that have raised money for projects such as launch services and image collection. While some of these involve the ISS, few use the ISS as a laboratory, as it was envisioned.

To increase the VC community's interest in the ISS, we recommend that NASA and CASIS continue to address the issues of information that preclude consideration of the ISS as a tool for applied, commercial research. In addition to the sort of outreach described earlier, providing and directing commercial users to information such as time schedules, pricing, equipment availability, and a simple checklist of the steps required to get a project on the ISS would render the facility much more accessible.[169] Commented one stakeholder cited in the 2012 CASIS report on Bioscience research at the ISS, "I don't see why we would want to do our experiments in space; we have perfectly good mice models on Earth."[170] It is always easier to say "no" to a new idea than "yes." CASIS and NASA must provide compelling reasons for entrepreneurs and their VC investors to do research and manufacturing on the ISS.

---

168 For relevant discussions see: David Lynn Hoffman and Nina Radojevich-Kelley. "Analysis of Accelerator Companies: An Exploratory Case Study of their Programs, Processes, and Early Results." *Small Business Institute Journal* 8, no. 2 (2012): 54–70.; Benjamin L. Hallen, Christopher B. Bingham, and Susan Cohen. "Do Accelerators Accelerate? A Study of Venture Accelerators as a Path to Success?" *Academy of Management Proceedings* (January, 2014).; and Daniel C. Fehder and Yael V. Hochberg. "Accelerators and the Regional Supply of Venture Capital Investment." Working Paper (September 19, 2014).

169 As described by Link and Maskin in an earlier chapter of this book.

170 CASIS. Maximizing the Value of the CASIS Platform–Biosciences Opportunity Map, 2012.

NASA and CASIS must also be realistic when considering the impact of the ISS's longevity on entrepreneurial interest. A company will be less eager to base its business model on a product that requires microgravity manufacturing if there is a strong chance that the manufacturing facility will be unavailable in a dozen years.

In the short term, though, we believe that NASA and CASIS could arouse interest in the VC community if it promoted its successes, facilitated company development through current and future angels/accelerators, and possibly provided financing to defray the costs of trials on the ISS. With information and access, the ISS could very likely attract some interest from VC-backed entrepreneurs. With the current fascination among entrepreneurs and venture capitalists for LEO opportunities, a few successes based on microgravity research may very well inspire another land rush, but this time, to space on the ISS.

Appendix A

VC Involvement in U.S. LEO Sector as of January 20, 2015 (Excluding Round Extensions)[a]

Note: While every effort was taken to identify all U.S. VC investments in LEO to date, certain investments may not have been identified given little public disclosure. Other startups may have been missed due to difficulty in identifying their involvement in LEO. Explanations of ambiguities within certain deals are described in footnotes.

Investment Details

COMPANY NAME (YEAR FOUNDED)	INVESTMENT DATE(S)	INVESTORS	AMOUNT INVESTED (U.S.$M)	SUBSECTOR AROUND TIME OF INVESTMENT
Orbital Sciences Corporation (1982)[b]	1983	Rothschild Inc.; Brentwood Associates; Northwest Bank	2.0	Transfer Orbit Stage (TOS) vehicles. Began investigating LEO in 1987.
SPACEHAB (now Astrotech) (1983)	1987	Poly Ventures; BEA Associates, Other undisclosed investors	1.8	Commercial space services (laboratory modules/cargo carries)
Constellation Communications (1991)[c]	1995/1996 Sep. 1997	SpaceVest (renamed Redshift Ventures) SpaceVest	2.5 2.5	Telecommunications satellites
Iridium Satellite, Inc. (1990s)[d]	2000	Syncom Venture Partners, Other undisclosed investors	25.0	Telecommunications satellites
SkyBitz (1993)	July 2000 Nov. 2002 Jan. 2004 Feb. 2007	Zero Gravity Venture Partners AIG Highstar Capital, Cordova Ventures AIG Highstar Capital; Cordova Ventures; Inverness Capital Partners; Motorola Ventures CIBC Capital Partners, AIG Highstar Capital, Inverness Capital Partners, Cordova Ventures, and Motorola Ventures	4.0 18.0 16.0 10.0	Asset tracking solutions operating over LEO and geostationary satellite systems
CartaSite (2004)[e]	March 2007	Canterbury Partners; Acadia Woods Partners	4.35	Remote monitoring for fleet management using LEO satellites
ORBCOMM (1993)[f]	Feb. 2004 Jan. 2006	SES Global, S.A.; Ridgewood Satellite LLC; OHB Technology A.G.; Northwood Capital Partners, others PCG Capital Partners; MH Equity Investors; Northwood Ventures; OHB Technology AG; Ridgewood Capital; Torch Hill Capital	26.0 110.0	Telecommunication satellites
Space Adventures (1998)	Feb. 2006	Prodea Systems	N/A	Space tourism (including to the ISS)

COMPANY NAME (YEAR FOUNDED)	INVESTMENT DATE(S)	INVESTORS	AMOUNT INVESTED (U.S. $M)	SUBSECTOR AROUND TIME OF INVESTMENT
Xcor Aerospace[g] (1999)	Aug. 2008 May 2014	Desert Sky Holdings Space Expedition Corporation; Space Angels Network, Esther Dyson	N/A 14.2	Reusable launch vehicle and propulsion systems
SpaceX[h] (2002)	Aug. 2008 Aug. 2009 Nov. 2010 Dec. 2012 Jan. 2015	Founders Fund Draper Fisher Jurvetson; DFJ Growth; Scott Banister Draper Fisher Jurvetson; Founders Fund; Musket Research Associates; Other undisclosed investors Draper Fisher Jurvetson; Rothenberg Ventures Google; Fidelity Investment	20.0 30.4 50.2 30.0 1,000.0	Rocket designer, manufacturer, and launcher.
Skybox Imaging (2009)	July 2009 July 2010 April 2012	Khosla Ventures Bessemer Venture Partners; Draper Associates; Khosla Ventures Asset Management Ventures; Bessemer Venture Partners; Canaan Partners; Khosla Ventures; Norwest Venture Partners; CrunchFund.	3.0 18.0 70.0	Microsat developer for earth imaging
Spire (formerly Nanosatisfi) (2012)	Feb. 2013 July 2014	Beamonte Investments; E-Merge; Lemnos Labs; Shasta Ventures; Fresco Capital RRE Ventures; Mitsui & Co Global Investment; Moose Capital; Qihoo; E-Merge; Promus Ventures	1.2 25.0	Nanosat developer (remote sensing) for data solutions
Planet Labs (formerly Cosmogia) (2010)	June 2013 Dec. 2013 Jan. 2015	Capricorn Management; Data Collective; Draper Fisher Jurvetson; First Round Capital; Founders Fund; Innovation Endeavors; O'Reilly AlphaTech Ventures Yuri Milner; AME Cloud Ventures; Capricorn Management; Data Collective; Draper Fisher Jurvetson; Felicis Ventures; First Round Capital; Founders Fund; Industry Ventures; Innovation Endeavors; Lux Capital Management; O'Reilly AlphaTech Ventures; Ray Rothrock Data Collective; AME Cloud Ventures; Capricorn Management; Draper Fisher Jurvetson; Felicis Ventures; First Round Capital; Founders Fund; Industry Ventures; Innovation Endeavors; Lux Capital Management; O'Reilly Alphatech Ventures; Ray Rothrock; Yuri Milner	13.1 52.0 70.0	Nanosat developer for earth imaging
NanoRacks (2009)	June 2013	E-Merge; Other undisclosed investors	2.6	ISS research platforms and payload coordination

COMPANY NAME (YEAR FOUNDED)	INVESTMENT DATE(S)	INVESTORS	AMOUNT INVESTED (U.S.$M)	SUBSECTOR AROUND TIME OF INVESTMENT
Planetary Resources (formerly Arkyd Astronautics) (2010)[i]	Aug. 2012 July 2013	I2BF Global Ventures I2BF; Dylan Taylor; Space Angels Network	N/A ("Less than tens of millions of dollars")[i] 1.5	Asteroid mining
Rocket Lab (2007)[k]	Oct. 2013	Khosla Ventures	N/A	Miniature satellite launch system
Acme Advanced Materials (2014)	Feb. 2014	Cottonwood Technology Fund; Pangaea Ventures	N/A	Materials manufacturer in microgravity
Accion Systems (2014)	Jan. 2015	Founders Fund; Founder Collective; Galvanize Ventures; GettyLab; RRE Ventures; SDF Ventures; Slow Ventures; TechU Angels; Other individual investors	2.0	Propulsion for small satellites

a   Data obtained from press releases, news articles, company websites, industry reports, CrunchBase, Preqin, and Thomson ONE.

b   Orbital Sciences Corporation also raised $50 million from LPs to finance the TOS vehicle in 1984/85.

c   While only the September 1997 has explicitly been made public, media releases in September 1997 noted that, "SpaceVest originally invested $2.5 million in the 6-year-old company [Constellation Communications] last year and the year before. Now the fund supplied another $2.5 million." While the language is slightly ambiguous, we interpreted this as a single investment of $2.5 million occurring in two tranches in 1995/96. See Bob Starzynski. "SpaceVest Adds $2.5M to Constellation Coffers." *Bizjournals.com* (September 22, 1997).

d   This investment was atypical in the sense the Syncom and other investors acquired the entirety of Motorola-backed Iridium, which declared bankruptcy in 1999. The investors also added another $130 million in commitments to recapitalize Iridium. In September 2009, investment bank Greenhill & Company (GHL) acquired Iridium for $560 million. For more information, see Frank McCoy. "The Art of Turning $25 Million Into $560 Million." *Theroot.com* (January 25, 2010).

e   CartaSite secured seed financing of $1.2 million in 2004, though no investors were disclosed. We presume the financing came from individuals. There is conflicting data as to whether CartaSite also received a $3.2 million investment in April 2006.

f   ORBCOMM also received investments from Silver Canyon Group and Centripetal Capital Partners, though it is unclear as to whether these investments were separate from or part of the deals listed in Appendix A.

g   Xcor Aerospace reported additional funding rounds of an undisclosed amount in June 2007 and $5 million in February 2012. We note, however, that the June 2007 investment appears to be solely from an angel network (Boston Harbor Angels). See, Xcor Aerospace. "Boston Harbor Angels Invests in XCOR Aerospace: Investment Fuels Expansion into New Markets." (June 7, 2007). The February 2012 investment was reportedly from angel investors and former venture capitalists, not traditional VC firms. See, Xcor Aerospace. "XCOR Aerospace Closes $5 Million Round of Investment Capital." (February 27, 2012). While we cannot be sure if other institutional investors were involved, the majority of funding seemed to be from individual investors/angel networks. We therefore do not include these as VC funding. We also note that Xcor's Series B round was not from a traditional VC group, but Space Expedition Corporation appears to have taken a minority stake and two of its officers joined Xcor's board. Xcor later acquired all operational subsidiaries of SXC in June 2014, while the parent SXC company remained a "passive investment holding company." See, Jeff Foust. "XCOR Acquires One of its Investors' Subsidiaries (Updated)," *NewSpace Journal* (July 1, 2014).

h   While reports had circulated that SpaceX raised an additional $200 million in August 2014, SpaceX representatives have refuted these claims. See Alan Ohnsman. "Musk's SpaceX Denies Blog Report of Capital Raising Plan." *Bloomberg Businessweek* (August 19, 2014).

i   Planetary Resources' financing history is not entirely transparent. We find it unclear as to whether the firm had two separate financing transactions in July 2013, as the company reportedly raised $1.5 million in a crowdfunding effort, as well as $1.5 million in "seed" funding during the same month from the investors noted in Appendix A.

j   "Russia-linked Company Invests in Asteroid Mining." *Moscow Times* (August 17, 2012).

k   We note that Rocket Lab is a U.S. corporation, but is based in New Zealand. For details, see Doug Messier. "A Closer Look at Rocket Labs' Technical Development." *Parabolicarc. com* (August 1, 2014).

CHAPTER 5

Directing vs. Facilitating the Economic Development of Low Earth Orbit

Mariana Mazzucato[1]
Douglas K. R. Robinson[1,2,3]

THIS CHAPTER FOCUSES on the challenges and opportunities associated with the shift away from NASA-centric development in low Earth orbit (LEO) toward an ecosystem with a mix of private, not-for-profit, and public actors. We focus on the question of whether NASA can and should direct the commercialization of LEO or whether its role should primarily be that of a facilitator. NASA has historically been the central definer of national space activities as a mission-oriented public agency with a clear ambition to direct innovation, not just to facilitate it. NASA's role in LEO, however, is in the process of changing. We describe how public-private partnerships have created both tensions and opportunities for NASA. We consider how these tensions and opportunities may play out between now and the projected end of the International Space Station (ISS) and in terms of how such relationships can be built to support NASA in its mission to explore and develop the rest of the solar system beyond LEO.

---

1 Science Policy Research Unit (SPRU), University of Sussex Brighton BN1 9QE.

2 Université Paris-Est, Laboratoire Interdisciplinaire Sciences, Innovations, Société (LISIS), ESIEE, F 77454 Marne-La-Vallée, France.

3 TEQNODE Limited, 282 rue Saint Jacques, 75005 Paris, France.

Disclaimer: The views and opinions of the authors do not necessarily state or reflect those of the U.S. Government or NASA.

Section 1. Directing Change Through Mission-Oriented Innovation Policy

Innovation policy is often justified through the need to correct different types of market failures. Market failures are defined as situations in which positive or negative externalities require corrections to help the private sector invest more (e.g., in public good areas such as R&D) or less (e.g., in activities that create pollution). Indeed, public goods such as basic science, which have high spillovers, are a typical example of an area with strong positive externalities. However, such market failure corrections, while important, do not describe the depth and breadth of the visible hand of the state in countries that have achieved smart innovation-led growth. In places like Silicon Valley, government policy has actively shaped and created markets, not only "fixed" them (Mazzucato 2013; 2015; 2016). Sectors including biotechnology, nanotechnology, and indeed the information technology (IT) revolution have been dynamic outcomes of active public policies—by agencies like NASA in civilian space activities, DARPA in the Department of Defense, and the National Institutes of Health in the Department of Health and Human Services—which have created and shaped markets, not only fixed them.

Such agencies have been driven by mission-oriented policies aimed at clearly defined, unambiguous technical goals (Foray et al. 2012). Such policies have often been active and "vertical,"[4] explicitly choosing concrete technologies, sectors, and even firms to support. Rather than assuming that the market will direct change, with "horizontal"[5] policies aimed only at the underlying competitive framework, active mission-oriented policies have been determining not only the rate of innovation but also its direction (Stirling 2009). Indeed, mission-oriented policies differ from purely sector-oriented policies as they present concrete problems for many different sectors to work toward. Going to the moon required innovation in robotics, spacecraft, textiles, material science, information technology, and many other areas.

As mission-oriented goals have tended to focus on grand technological objectives, often related to security issues, the justification for such missions in the United States has changed over time. While military motives dominated in the 1950s and 1960s, the aim since the 1970s has been to improve economic and competitive positions,

---

[4] *Vertical approaches* to innovation policy focus on direct investments (in both basic and applied areas) on specific technologies and sectors (e.g., biotechnology, nanotechnology, clean-technology), and financing specific firms through public venture capital or public loans. Vertical policies determine not only the occurrence of innovation but also guide direction—determining the boundaries within which private sector innovation and experimentation can happen.

[5] *Horizontal approaches* to innovation policy focus on (a) creating the framework conditions for innovation through the background (necessary) conditions (e.g., education, science-industry links, fundamental research and infrastructures) and (b) supporting innovation in the private sector through indirect measures such as tax incentives. The horizontal approach tends to rely on the market to decide the direction of change.

and innovation policy has indeed extended to many fields, including health and energy. This has coincided with the explicitly stated objective for innovation policy to aim at allowing public funding to have clear commercial outcomes. The priority for commercialization has affected the goals of government research funding, causing agencies like DARPA, NIH, and NASA to justify success of research by proving or providing a convincing argument for future economic value of their science and technology bases (Weiss 2014).

NASA's initial mission-oriented programs for innovation were driven by security and maintaining technical leadership over other nations, with the Apollo program born to compete with the Soviet program in the "race for space." In achieving these missions, NASA's public funding was linked to different actors in the U.S. innovation system, including universities and private sector actors (through procurement), with NASA retaining the central (and vertical) directing role. The post-1970s emphasis on commercialization has been shifting this ecosystem, not only with the effect of space on the economy often seen as just as important as the effect of space on security, but also in terms of NASA's position in the innovation system. In recent years, more horizontal, less vertical, measures are being used to facilitate commercialization, with NASA still playing a crucial role, but allowing the private sector to take on a more directional role.

The shift toward making commercialization more prominent in NASA's missions is particularly visible with the congressional legislation that directed NASA to competitively select a not-for-profit organization to enhance use (public and private) of the International Space Station (ISS). This is also evident in the 2013 National Space Transportation Policy, which has called for a greater use of private space transportation to low Earth orbit (LEO).[6] In addition to the recent activities supporting the development of additional private capabilities to access LEO, and the active support of nongovernmental users of the U.S. National Lab aboard the International Space Station, LEO is becoming increasingly populated by a variety of American firms, public research organizations, and other actors.

NASA's inclusion of commercialization objectives in its LEO activities, as well as its technological goals, has created different types of challenges, both associated with a type of programmatic broadening.

Firstly, it is more difficult to evaluate commercialization than technological goals, as the former's success is dependent on activities additional to those of NASA and its partners, most notably the market and its framing conditions. These multiple dependencies for successful commercialization make it difficult to predict the timeframes

---

6 See FY 2014 Annual Performance Report and FY 2016 Annual Performance Plan for more details. Also, see the H.R.4412 – National Aeronautics and Space Administration Authorization Act of 2014 for details of the NASA mission: *https://www.congress.gov/bill/113th-congress/house-bill/4412*; National Space Transportation Policy 2013 can be found on the White House website: *https://www.whitehouse.gov/sites/default/files/microsites/ostp/national_space_transportation_policy_11212013.pdf*.

for successful commercialization. This open-endedness differs greatly from the centralized planning of large technological development projects such as Apollo.

Secondly, the move towards commercialization as an objective has, through NASA's policies, greatly increased the number of stakeholders in LEO, which has changed the number and make-up of the constellation of actors that interact in the emerging LEO economic system. This is visible already and may broaden further in the years to come.

This chapter focuses on how NASA's aim of developing a sustainable LEO ecosystem, is affected by these two types of broadening. We argue that the fate of LEO, and also the ability of NASA to fund future exploration missions, is dependent on (1) the success of the Earth-to-LEO innovation system, (2) the associated innovation policy mix, and (3) how this is linked to NASA's key objectives as a national mission-oriented public agency.

The chapter introduces key concepts that allow us to describe the innovation system in space in which NASA has played a key position. These concepts are (a) a "systems of innovation" perspective on innovation policy; (b) the notion of "mission-oriented" policies as shaping and creating markets (not just fixing them, as in the traditional market failure perspective); and (c) the distinction between vertical policies that set the direction of change versus more horizontal policies that assume the market will determine the direction with the state only facilitating it. We conclude by considering the effect of current and emerging relationships on the ability of NASA to play a mission-oriented role in LEO in the future.

Section 2. Systems of Innovation: Creating and Shaping Markets

2.1 Mission-Oriented Policies: From "Fixing" to Creating Markets

Market failure theory justifies public intervention in the economy only if it is geared toward fixing situations in which markets fail to efficiently allocate resources (Arrow 1951). The market failure approach suggests that governments intervene to "fix" markets by investing in areas with public goods characteristics (such as basic research or drugs with little market potential) and by devising market mechanisms to internalize external costs (such as pollution) or external benefits (such as herd immunity).

Market failure has often been seen as a necessary but not sufficient condition for governmental intervention (Wolf 1988). This has resulted from a view that the gains of intervention are often outweighed by the associated costs due to governmental failures—such as capture by private interests (e.g. nepotism, cronyism, corruption, rent-seeking), misallocation of resources (e.g., "picking losers"), or undue competition with private initiatives ("crowding out") (Tullock et al. 2002), (Krueger 1974), (Falck et al. 2011), (Friedman 1979). Thus, there is a trade-off between two inefficient outcomes; one is generated by free markets (market failure) and the other by governmental intervention (government failure). The solutions advocated by Neo-Keynesians focus on correcting failures such as imperfect information (Stiglitz and Weiss 1981). Solutions advocated by public choice scholars (Buchanan 2003) focus on leaving resource allocation to markets (which may be able to correct their failures on their own).

While market failure theory provides interesting insights, it is at best useful for describing a steady state scenario in which public policy aims to put patches on existing trajectories provided by markets (Mazzucato 2016). It is less useful when policy is needed to dynamically create and shape new markets. This means it is problematic for addressing innovation and societal challenges because it cannot explain the kinds of transformative, catalytic, mission-oriented public investments that in the past have created new technologies and sectors which did not exist before (the Internet, spaceflight, nanotech, biotech, clean-tech), and whose disruptive nature many incumbent private sector actors ignored or feared.

To understand the advent of such technological revolutions, innovation scholars have thus emphasized the role of policies that have actively created markets, not just fixed them, through mission-oriented objectives (Mowery 2012; Mazzucato and Penna 2015). NASA has often been characterized as the prototypical mission-oriented agency, based on the Apollo Program's mission of landing a man safely on the Moon and returning him safely to the Earth. It was such mission-oriented investments that coordinated public and private initiatives, built new networks, and drove the entire techno-economic process, which resulted in the creation of new markets. Indeed, all the technologies that make modern cell phones "smart" were financed by such mission-oriented governmental programs (Internet, GPS, touchscreen, and voice activated SIRI technology) (Mazzucato 2013).

The mission-oriented literature contains many useful empirical studies, such as analysis of different technology policy initiatives in the United States (Chiang 1991; Mowery et al. 2010), in France (Foray 2003), in the United Kingdom (Mowery et al. 2010), and in Germany (Cantner and Pyka 2001); and studies of mission-oriented agencies and policy programs, including military R&D programs (Mowery 2010), the National Institutes of Health (Sampat 2012), grand missions of agricultural innovation in the United States (Wright 2012), and energy (Anadón 2012), among others. While mission-oriented programs are intrinsically dynamic, with feedback loops between missions and achievements, the tools used to evaluate such public policies have remained static, coming from the market failure theory toolbox (despite the fact that many studies draw on the dynamic innovation systems perspective from evolutionary economics).

The mission-oriented framework (Ergas 1987; Freeman 1996; Mowery 2010) helps understand why it is that public sector funds have been necessary not just for classical public good areas like basic research but investments along the entire innovation chain. **Figure 5.1** illustrates the variety of public organizations (in bold) that

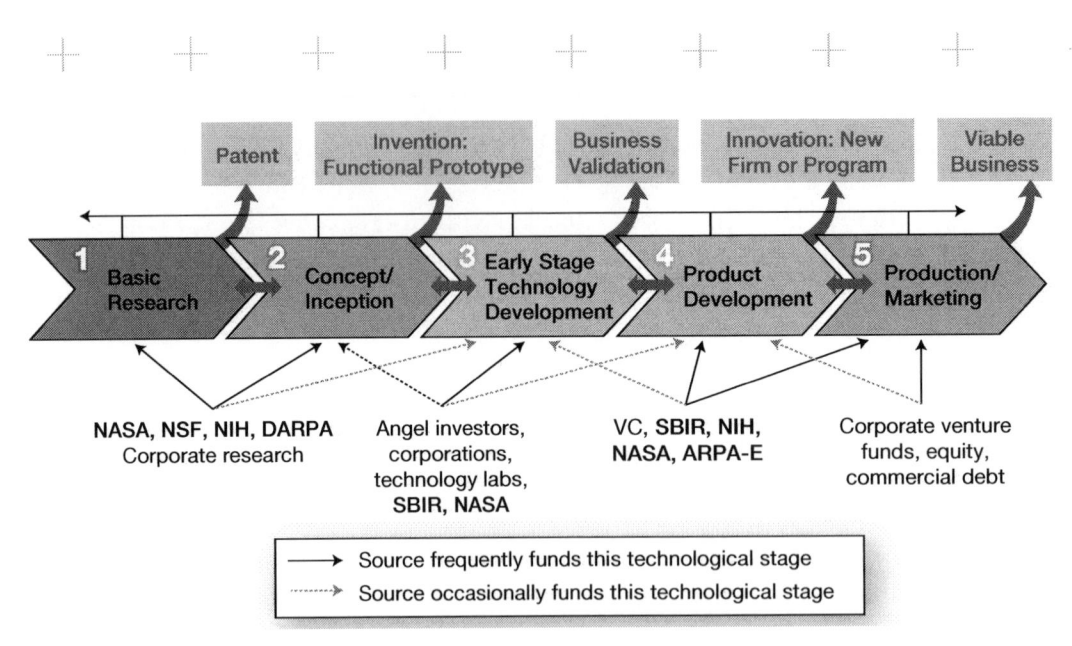

FIGURE 5.1: Public and Private Investments Along Entire Innovation Chain (Source: Authors' addition of public agencies to underlying figure by Auerswald and Branscomb 2003.)

have been active along the innovation chain. This includes basic research by agencies such as the National Science Foundation (NSF); applied research by NASA, DARPA in the Department of Defense, ARPA-E in the Department of Energy, and even early-stage seed financing for companies, through agencies and programs like the Small Business Innovation Research (SBIR) program. Both ARPA-E and DARPA have managed to attract high-level experts (from universities, and the private sector), through the use of 4–5 year secondments, with the objective of stimulating innovation in sustainable/renewable energy technologies (Bonvillian et al. 2011). The DOE has used the Agreements for Commercializing Technology (ACT) to promote and grow the commercialization of its R&D to create new products for the market (Epstein 2012; Jaffe and Lerner 2012).

Investments by such organizations have often been guided by procurement; this stimulates both the supply side and the demand side, with the demand side being key to creating a market for new technologies (Mazzucato 2015). The creation of new markets is often an outcome of mission-oriented programs. Public agencies have taken up this role due to the hesitation of a risk-averse private sector, particularly concerning breakthrough technology fields and new markets.

Increasingly such mission-oriented investments have been found to be key for allowing innovation to take off in a way that generates long-term growth, and understanding the history of mission-oriented policies has been crucial for thinking about new policies needed to address "grand societal challenges" (Foray et al. 2012). In historical mission-oriented R&D projects of the past, such as the Manhattan and Apollo programs, all funding has been provided by public U.S. Federal agencies, but for current societal challenges and LEO commercialization specifically publicly funded R&D, although vital, will be only one of a number of sources of R&D

investment. In recent years, there have been calls for a return to such mission-oriented policies, extended to areas that can address grand societal challenges ranging from the aging demographic problem being faced by Western nations to the global challenges concerning climate change (Foray et al. 2012). However, grand societal challenges concern the socioeconomic system as a whole, which often implies large-scale transformations with multiple actors and elements (Kuhlmann and Rip 2014; Geels 2004). This is in stark contrast to the missions of the past, which were mainly technical and more vertical in their solutions (Foray et al. 2012).

Understanding the role of new actors required to confront missions that are socioeconomic and not just technical, requires a "system of innovation" viewpoint.

2.2 From Market Failures to System Failures

Systems of innovation (whether sectoral, regional, or national) embody dynamic links between various innovation actors and institutions (firms, financial institutions, research/education, public sector funds, and intermediary institutions), as well as links within organizations and institutions (Freeman 1995). The "systems of innovation" approach (Freeman 1995) to understanding innovation policy, provides key insights into not only the limits of market failure theory in terms of justifying the depth of investments that have been necessary for the emergence of radical technological change, but also the breadth of the different actors involved. Innovation policy has historically taken the shape of measures that (1) support basic research, (2) aim to develop and diffuse general-purpose technologies, (3) develop certain economic sectors that are crucial for innovation, and (4) promote infrastructural development (Freeman and Soete 1997).

By highlighting the strong uncertainty underlying technological innovation, as well as the very strong feedback effects that exist between innovation, growth, and market structure, the systems of innovation view emphasizes the "systems" component of technological progress and growth.[7] Systems of innovation have been defined as "the network of institutions in the public and private sectors whose activities and interactions initiate, import, modify and diffuse new technologies" (Freeman 1995), or "the elements and relationships which interact in the production, diffusion and use of new, and economically useful, knowledge" (Lundvall 1992, p. 2).

The emphasis here is not on the stock of R&D, but on the circulation of knowledge and its diffusion throughout the economy. Institutional change is not assessed

---

7 The emphasis on heterogeneity and multiple equilibria requires this branch of theory to rely less on assumptions of representative agents (the average company) and unique equilibria, which remain central to mainstream (neoclassical) economics. Rather than using incremental calculus from Newtonian physics, mathematics from biology (such as distance from mean replicator dynamics) are used, which can explicitly take into account heterogeneity and the possibility of path dependency and multiple equilibria. See Mazzucato (2000).

through criteria based on static allocative efficiency, but rather on how it promotes technological and structural change. Individual firms are seen as part of a broader network of firms with whom they cooperate and compete. The system of innovation can be interfirm, regional, national, or global. From this perspective it is the network—not the firm— that is the unit of analysis. The network consists of customers, subcontractors, infrastructure, suppliers, competencies, or functions and the links or relationships between them. The competencies that generate innovation are part of a collective activity occurring through a network of actors and their links or relationships (Freeman 1995).

The causation in the steps between basic science, large-scale R&D, applications, and finally to diffusing innovations is not linear. Instead, innovation networks are full of feedback loops between markets and technology, applications, and science. In the linear model, the R&D system is seen as the main source of innovation, reinforcing economists' use of R&D statistics to understand growth. In this more nonlinear view, the roles of education, training, design, quality control, and effective demand are just as important.

Furthermore, the serendipity and uncertainty that characterizes the innovation process is useful in terms of understanding the rise and fall of various economic powers in history. For example, it explains the rise of Germany as a major economic power in the 19th century as a result of state-fostered technological education and training systems. It also explains the rise of the United States as a major economic power in the 20th century as a result of the rise of mass production and in-house R&D. The United States and Germany became economic powers for different reasons, but they both paid attention to developing systems of innovation rather than focusing narrowly on raising or lowering R&D expenditures.

2.3 Setting the Direction of Change vs. (Just) Facilitating It

Investments in innovation involve choices regarding which innovations or sectors to invest in and to what extent. This choice between options and what amount of resources to invest means that innovation has both a *rate* and *direction* (Stirling 2009; Smith et al. 2005; Robinson and Propp 2008), where the rate is dependent on the intensity of resources invested and the direction is guided by choices made based on shared visions, goals, and policies. Articulating and enacting such direction is a key part of mission-oriented institutions (Mowery et al. 2010; Foray et al. 2012; Sampat 2012; Edquist et al. 2012). In the United States, many mission-oriented public agencies have missions to catalyze innovation (increase the rate) in line with their missions (direction). Examples include DARPA in the Department of Defense, NIH in the Department of Health and Human Services, ARPA-E in the Department of Energy, and NASA, all of which have had profound impacts on economic growth and, in many cases, the creation of new markets.[8] Directionality brings the focus away from the narrow discussion of whether or not to "pick winners" to a broader discussion of how the picking should occur (Mazzucato 2015; 2016). Setting the direction is not about choosing a narrow set of sectors, but choosing the problems and the missions for a broad group of sectors to react to. Furthermore, in order for public agencies (such as DARPA or NASA) to be able to direct such missions it is essential for them to attract the kind of scientific and technological expertise which will allow them to do so. Evidence suggests that it is precisely in mission-oriented organizations that high-level experts (natural and social scientists) will find it an honor to work in the public sector, even if the monetary compensation is lower. When instead the mission of a public agency is simply to fix market failures, it becomes much harder to make it attractive to work in such agencies, creating a self-fulfilling prophecy whereby the ability to direct changes becomes more difficult, and bureaucrats are expected to have knowledge they don't have—so they get accused of being unable to pick winners.

Systems of innovation require different types of policies. Vertical policies have been more directional and active, focusing on directing change. Horizontal policies have been more focused on the background conditions necessary for innovation, allowing the direction to be set by the private sector.

While both horizontal and vertical policies are required, it can be said that horizontal policies are more about facilitating innovation in the private sector, while vertical policies embody a more active role for the public sector in directing change not only facilitating it, often through missions which require actively creating and shaping markets—not only fixing them (Mazzucato 2015).

Section 3. Three Questions for NASA's Future Innovation Policy Mix

This collection of essays was initiated due to an interest in stimulating a sustainable LEO-ecosystem built on public-private partnerships (PPPs) with fully public and fully private activities as part of the mix. NASA's portfolio of PPPs has shown a transition from a vertical innovation policy to a more distributed innovation policy, where goals are set by multiple actors with different criteria of success and directions of development. When considering this new constellation of actors and relationships, what can be said about the state of the LEO innovation system and NASA's role within it?

We approach this through a number of challenges that we have seen emerging in the discussion around LEO commercialization. These challenges can be grouped into three broad questions about the factors and actors shaping the destiny of the LEO ecosystem:

---

8 For some very visible examples: NASA and the Apollo program, NIH and the war on cancer, DARPA and Information Technology.

1. **Directionality:** Is the increasing emphasis on commercialization affecting NASA's mission-oriented innovation policy in LEO? In particular, who is directing change?

2. **Risks and Rewards:** How are current public–private partnerships affecting the balance of risk and reward in the conduct of spaceflight in LEO?

3. **Organizational Capacity:** Is the transition to PPPs reducing innovation capacity within NASA itself? Is there a danger that the internal capacity to address innovation challenges will be reduced for future missions?

3.1 Directionality

The central position of NASA in the space "system of innovation" has meant that, for more than 50 years, NASA has directly financed technological innovation to achieve its missions, setting the directions of change and overseeing the private sector companies that have been contracted to deliver the technologies. Today, the missions of technological innovation in LEO are being broadened to include commercialization objectives. NASA is attempting to create new markets that fuel a sustainable Earth-LEO economy, or, as Sam Scimemi, director of the ISS Program at NASA HQ, put it, to "sustained economic activity in LEO enabled by human spaceflight, driven by private investments, creating value through commercial supply and demand" where the "destiny of LEO beyond ISS is in the hands of private industry outside the government box."[9]

While commercialization did occur before as a spillover of NASA missions, commercialization is now more central to NASA's mission in LEO in general. An example can be seen in the Commercial Space Act of 1998:

> The Congress declares that a priority goal of constructing the International Space Station is the economic development of Earth orbital space. The Congress further declares that free and competitive markets create the most efficient conditions for promoting economic development, and should therefore govern the economic development of Earth orbital space. The Congress further declares that the use of free market principles in operating, servicing, allocating the use of, and adding capabilities to the Space Station, and the resulting fullest possible engagement of commercial providers and participation of commercial users, will reduce Space Station operational costs for all partners and the Federal Government's share of the United States' burden to fund operations.[10]

---

9 Presentation given by Sam Scimemi at NASA Headquarters Washington, DC, December 10, 2014; http://www.nasa.gov/sites/default/files/files/NASA_Sam_Scimemi.pdf.

10 The Commercial Space Act of 1998 (Public Law 105-303). http://www.nasa.gov/offices/ogc/commercial/CommercialSpaceActof1998.html.

The process of LEO commercialization began as part of NASA's human spaceflight activities and its in-orbit operations in the early 1980s, during the early stages of the Shuttle program, but was part of a NASA-centered space ecosystem based on direct NASA oversight of the private sector. In recent years, the innovation system has shifted to include more actors, and devolution of the management of a large share of research and innovation activities on the ISS to private actors and intermediaries.

This shift regarding mission-oriented policy related to human spaceflight and in-orbit operations means that NASA has to handle two forms of broadening, mentioned in our Introduction: a broadening of mission goals to include goals that are more economic in focus and less technological, and a broadening of the number and types of actors in the innovation system. Consequently, unambiguous objectives with a clearly defined endgame are more difficult to define in the emerging LEO ecosystem, which involves multiple actors with different motivations as well as a decentralization of power in terms of shaping and directing the ecosystem.

The degree of decentralization is different for various parts of the ecosystem. What is clear is that the private sector is becoming a prominent force in the directionality of the LEO ecosystem. A new set of actors that has emerged over the past six years is brokers. Brokers in LEO are intermediaries between suppliers and users of LEO facilities for the purpose of research and innovation. Brokers navigate the complex LEO ecosystem in order to connect potential users of the ISS with the Government. There are two types of brokers for the ISS: the Center for the Advancement of Science in Space (CASIS), which is a not-for-profit organization set up to promote scientific research on the US National Lab by soliciting potential users; and NanoRacks, a Texas-based company, founded in 2009. CASIS connects existing and emerging firms with value chains in areas such as the pharmaceutical industry and advanced materials with the facilities of the ISS National Lab. NanoRacks connects commercial users with novel uses of existing (or slightly modified) capabilities, as illustrated by the small satellite deployer, which has created a new market with only minor developments in technical capability.

Another form of decentralization was the use of Space Act Agreements in executing the Commercial Orbital Transportation Services Program (COTS) and the Commercial Crew Development Program (CCDev), which meant more control of design and development shifted to the contracted firm (e.g., the key milestones and the associated price were defined by the private contractor, which meant they delivered on time or did not get paid).

Although it is too early to say what the ultimate outcome will be, NASA is undeniably delegating some of its power in directing the development of LEO to the private sector.

3.2 Risks and Rewards

Commercial activities on the International Space Station U. S. National Laboratory are subsidized by U.S. taxpayers and brokered via CASIS. The argument put forward is that providing the ISS facility for free to the private sector would stimulate the creation of knowledge, trigger innovation, and contribute to national economic growth. From this perspective, one could argue that there are two overlapping and entangling systems. The first is one could be labeled the "LEO ecosystem," with activities focused on both exploration and exploitation of LEO. The second can be labeled the "U.S. National Innovation System," which is comprised of many sectors and value chains and carried by a more diverse constellation of actors and activities.

The policy challenge is how to connect the two systems to generate benefits for the U.S. taxpayer, as well as maintain a sustainable U.S. presence in LEO. In the current arrangement for LEO, one sees largely a horizontal policy of government-subsidized commercialization that could benefit the U.S. National Innovation System via the private sector actors who make use of the no-cost access to the ISS and create wealth broadly.

Creating a sustainable LEO ecosystem implies one of two scenarios. Scenario 1 involves closing the financial loop between public investment of resources (NASA) and private sector gain (subsidized users of the ISS) in a manner that reinvested gains from this process back into space infrastructure.

Scenario 2 involves closing the financial loop between public investment of resources (NASA) to private sector gain (subsidized users of the ISS), with the private sector investing in orbital facilities to which the public sector would have access.[11] Could joint ventures between the public and private sector be the way forward? Indeed, a number of our interviewees suggested that something between licensing and a COTS-type initiative to create a next-generation ISS-type facility could be an interesting approach after its success in orbital transportation.

The emphasis on the ISS's relevance to existing markets puts pressure on the access and support of potential breakthrough technologies that could create new markets. The case of Made In Space shows that alternative approaches to space activities are possible; in this case, with fabrication in orbit opening new possibilities for developing lightweight and fragile structures in space, which would be too delicate for launch. At same time is hard to see how Made In Space would have emerged as a viable company without the direct support of NASA Ames and later the SBIR program. This raises the question of what should be the optimum mix of approaches to create a sustainable LEO ecosystem.

Another issue is intellectual property. There is an ongoing discussion with the space policy community about the waiving of intellectual property rights and licensing fees to enhance commercialization. Should there be limited direct financial returns to

---

11 Other scenarios may be possible, mixing these two extremes.

NASA or CASIS for drug development that involves the ISS National Lab? If not a direct return to NASA or CASIS, what other deals might be possible to improve the returns to taxpayers?

3.3 Organizational Capacity

With the devolution of LEO commercialization responsibilities to the private sector, a key question for NASA as the Agency responsible for U.S. civilian space activities is whether or not NASA need maintain the ability or capacity to intervene and shape LEO activities if necessary. In the context of another mission-oriented agency, the Department of Defense, Mowery (2012) observed that, in addition to R&D expenditure supporting weapons development, the DOD often funds R&D in peacetime with the aim of making future weapons possible, and to provide knowledge to help it decide what kinds of weapons to try to develop and ultimately to procure and use.

It is important to consider whether and how transitioning spaceflight activities in LEO could potentially reduce the internal capacity of NASA for its future exploration missions. How would it affect NASA's ability to accumulate the kind of knowledge and expertise (and attract the talent needed) that were essential for the Apollo and Shuttle periods of U.S. space activities? Relying on the private sector for space transportation offers potential advantages in terms of improved cost and time to delivery, although it also means that the related knowledge and experience being accrued lies in the private domain. What does this mean for NASA's organizational capacity to absorb new knowledge and embark on further technology developments? And although it is clear that LEO commercialization can only happen once private sector companies are activity engaged in and investing in demand-side applications, such as those provided for by the microgravity environment, to what extent should NASA maintain and active and perhaps even leading role in the development of these applications? For private sector companies, this is referred to in terms of "absorptive capacity;" that is, the degree to which engaging in the actual process of R&D opens up your capacity to foresee, understand, and absorb new technological opportunities, even when these are unrelated to your own R&D (Cohen and Levinthal 1989). In sum, NASA should be careful to continue to nurture its own "absorptive capacity" if it wants to remain a mission-oriented innovation-based Agency.

Section 4. Conclusion

The narrative often visible in discussions of market creation and innovation describes the private sector as a major driver of innovation with the public sector only important in how it levels the playing field and creates the conditions for innovation to happen in the private sector. At best it is seen as a "market fixer" and at worst, an impediment to innovation. In this chapter we have argued that historically NASA has played an active role in innovation and market creation through its mission-oriented (vertical) policies, which have actively created new markets related to space, not just fixed them. Indeed this tallies with other studies, where public agencies have taken up this role due to the hesitation of a risk-averse private sector, particularly concerning breakthrough technology fields and new markets in the Internet, biotech, and nanotech sectors (Mazzucato 2013). Indeed, the ability to create new markets is a fundamental aspect of mission-oriented programs.

There has been a considerable shift from a fully NASA-directed U.S. presence in LEO, to a wider and more diverse ecosystem in LEO of public and private actors, interacting through a number of different forms of relationship, though the majority of activities are still heavily supported both financially and technically by NASA.

By placing a stronger emphasis on a "commercial approach" to LEO, NASA has delegated a large share of its ability to direct the LEO ecosystem and its link to the U.S. national innovation system to brokers and private actors. This represents a shifting of emphasis from mission-oriented vertical policies to distributed horizontal policies.

Successful innovation in different sectors has always required both vertical and horizontal policies, and hence this chapter provides a strong recommendation for NASA to continue in its path of doing both, rather than seeing horizontal, less "active," policies as the only way for commercialization to happen. We have outlined three challenges that should be considered (1) the issue of directionality, (2) the risk-reward balance for a sustainable link between LEO and the national innovation system and (3) NASA's organizational capacity to intervene and shape innovation in such a coupled LEO ecosystem and national innovation system.

Concerning the viability of a sustainable LEO ecosystem, our interviews revealed a concern about the availability of orbital facilities beyond 2024—there is a fair amount of uncertainty about whether there will be continued access to space on a comparable platform for further product development. Does a sustainable link between the LEO ecosystem and the national innovation system entail a need for next-generation orbital facilities and, if so, who should take the lead? Could it be something similar to the procurement-and-use approach demonstrated by the COTS program? Would the use of such a facility be brokered through agents such as those present today? How would this connect with NASA-supported breakthrough innovators such as Made In Space? These points raise issues in line with the challenges outlined regarding the mix between vertical and horizontal approaches and the risk-reward balance.

We have argued that the ability to create new markets is a fundamental aspect of mission-oriented programs. With the current transition towards a less mission-oriented (vertical) approach towards a more distributed (horizontal) approach, and the potential for ISS retirement in 2024, should an active directing role be foregrounded once again, as was clearly seen in the Apollo years?

To answer these questions going forward, NASA will face a number of challenges related to setting the directionality of change, building the organizations needed to do so, enabling a more dynamic evaluation of public investments, and achieving a mutualistic risk-reward relationship with the private sector.

References

Anadón, L. D. (2012). "Mission-oriented R&D institutions in energy between 2000 and 2010: A comparative analysis of China, the United Kingdom, and the United States." *Research Policy* 41 (10): 1742–1756.

Arrow, K. (1951). "An extension of the basic theorems of classical welfare economics." Paper presented at the Second Berkeley Symposium on Mathematical Statistics and Probability, Berkeley.

Auerswald, P. E., and L. M. Branscomb. (2003). Valleys of death and Darwinian seas: Financing the invention of innovation transition in the United States. *Journal of Technology Transfer* 28(3–4): 227–239.

Block, F. (2008). Swimming against the current: The rise of a hidden developmental state in the United States. *Politics & Society* 36(2): 169–206.

Block, F., and M. Keller. (2011). *State of innovation: The U.S. government's role in technology development.* Boulder: Paradigm.

Bonvillian, W. B., and R. Van Atta. (2011). ARPA-E and DARPA: Applying the DARPA model to energy innovation. *Journal of Technology Transfer* 36(5): 469–513.

Cantner, U. and A. Pyka. (2001). Classifying technology policy from an evolutionary perspective. *Research Policy* 30(5): 759–775.

CASIS (2013) The 2012 CASIS Annual Report. Available at *http://ww2.iss-casis.org/Casis_Annual_Report_2012*.

Chiang, J.-T. (1991). From "mission-oriented" to "diffusion-oriented" paradigm: The new trend of U.S. industrial technology policy. *Technovation* 11(6): 339–356.

Cohen, W. M., & Levinthal, D. A. (1989). Innovation and learning: the two faces of R & D. *The Economic Journal*, 99(397), 569–596.

Edler, J., and L. Georghiou. (2007). Public procurement and innovation—Resurrecting the demand side. *Research Policy* 36(7): 949–963.

Edquist, C. (2001). The Systems of Innovation Approach and Innovation Policy: An account of the state of the art. In DRUID Conference, Aalborg (pp. 12-15). Available at: *http://citeseerx.ist.psu.edu/viewdoc/download?doi=10.1.1.336.4438&rep=rep1&type=pdf*.

Edquist, C. and J. M. Zabala-Iturriagagoitia.(2012). Public procurement for innovation as mission-oriented innovation policy. *Research Policy* 41(10): 1757–1769.

Epstein, J. (2012). Panel V: Clustering around the lab—Best practices in federal laboratory commercialization. See *http://www.ncbi.nlm.nih.gov/books/NBK115038/?report=printable*.

Falck, O., C. Gollier, and L. Woessmann (2011). Arguments for and against policies to promote national champions. In O. Falck, C. Gollier, and L. Woessmann (eds.), *Industrial policy for national champions* (pp. 3–9). MIT Press.

Foray, D., D. C. Mowery, and R. R. Nelson (2012). Public R&D and social challenges: What lessons from mission R&D programs? *Research Policy* 41(10): 1697–1702.

Freeman, C., and L. Soete (1997 [1974]). *The economics of industrial innovation* (3rd ed.). MIT Press.

Friedman, B. M. (1979). "Crowding out or crowding in? The economic consequences of financing government deficits." *Brookings Papers on Economic Activity* 3: 593–654.

Geels, F. W. (2004). From sectoral systems of innovation to socio-technical systems: Insights about dynamics and change from sociology and institutional theory. *Research Policy* 33(6): 897–920.

Ginzburg, E., J. W. Kuhn, J. Schnee, and B. Yavitz. (1976). Economic impact of large public programs: The NASA experience. *http://ntrs.nasa.gov/archive/nasa/casi.ntrs.nasa.gov/19760016995.pdf.*

Jaffe, A. B., and J. Lerner (2001). Reinventing public R&D: Patent policy and the commercialization of national laboratory technologies. *Rand Journal of Economics*: 167–198.

Kuhlmann, S. and J. Edler. (2003). Scenarios of technology and innovation policies in Europe: investigating future governance. *Technological Forecasting and Social Change* 70(7): 619–637.

Kuhlmann, S. and A. Rip (2015). The challenge of addressing grand challenges. In: von Schomberg, R. (ed.): *The future of research and innovation*. European Commission.

Lerner, J. (1996). The government as venture capitalist: The long-run effects of the SBIR program (No. w5753). National Bureau of Economic Research.

Mazzucato, M. (2000). *Firm size, innovation, and market structure: the evolution of industry concentration and instability.* Edward Elgar Publishing.

Mazzucato, M. (2013). *The entrepreneurial state: Debunking the public vs. private myth in risk and innovation.* Anthem.

Mazzucato, M. (2015). "A mission-oriented approach to building the entrepreneurial state," Report commissioned by U.K. government. *https://www.gov.uk/government/news/long-term-growth-innovations-role-in-economic-success*

Mazzucato M. (2016) "From Market Fixing to Market-Creating: A new framework for innovation policy", Forthcoming in Special Issue of *Industry and Innovation*: "Innovation Policy–can it make a difference?" DOI 10.1080/13662716.1146124

Mazzucato, M. and C. Penna. (2015) *Mission-oriented finance for innovation: New ideas for investment-led growth.* Rowman & Littlefield.

Mowery, D. C. (2012). Defense-related R&D as a model for "grand challenges" technology policies. *Research Policy* 41(10): 1703–1715.

Mowery, D. C., R. R. Nelson, and B. R. Martin (2010). Technology policy and global warming: Why new policy models are needed (or why putting new wine in old bottles won't work). *Research Policy* 39(8): 1011–1023.

Mowery, D. C. (2010). Military R&D and innovation. In *Handbook of the economics of innovation* (vol. 2), B. H. Hall and N. Rosenberg (eds.), pp. 1219–1256.

NASA (1958). "National Aeronautics and Space Act of 1958," Public Law 85-568, as reprinted in *Exploring the unknown: Selected documents in the history of the U. S. civil space program, vol. 1., organizing for exploration,* John M. Logsden ed. NASA SP-4407, 1995.

NASA (2011). *Voyages: Charting the course for sustainable human space exploration.* NP-2011-06-395-LaRC.

NASA (2014). *Commercial orbital transportation services: A new era in spaceflight.* NASA/SP-2014-617.

National Aeronautics and Space Administration. *Extending the operational life of the International Space Station until 2024.* NASA Office of Inspector, 2014. *http://oig.nasa.gov/audits/reports/FY14/IG-14-031.pdf.*

Nelson, R. R., and N. Rosenberg. (1993). *Technical innovation and national systems. National innovation systems: A comparative analysis.* Oxford University Press.

Perez, C. (2002). *Technological revolutions and financial capital: The dynamics of bubbles and golden ages.* Edgar Elgar.

Pittman, B. and D. J. Rasky (2013). "Developing cislunar space using the COTS model." AD Astra Spring. *http://www.nasa.gov/sites/default/files/files/Adastra_2013_Pittman-Rasky.pdf.*

Polanyi, K. (2001 [1944]). *The great transformation: The political and economic origins of our time.* Boston: Beacon Press.

Pro-Orbis (2010). Reference model for the International Space Station U.S. National Laboratory. September 20.

Reinert, E. S. (2007). *How rich countries got rich and why poor countries stay poor.* Constable.

Report of the Advisory Committee on the Future of the U.S. Space Program, December 1990, *http://history.nasa.gov/augustine/racfup1.htm.*

Robinson, D. K. R., and T. Propp (2008). Multi-path mapping for alignment strategies in emerging science and technologies. *Technological Forecasting and Social Change* 75(4): 517–538.

Sampat, B. N. (2012). Mission-oriented biomedical research at the NIH. *Research Policy* 41: 1729–1741.

Smith, A., A. Stirling, and F. Berkhout (2005). The governance of sustainable socio-technical transitions. *Research Policy* 34(10): 1491–1510.

Soete, L., and A. Arundel (1993). *An integrated approach to European innovation and technology diffusion policy: A Maastricht memorandum.* Luxembourg: Commission of the European Communities, SPRINT Programme.

Stiglitz, J., and A. Weiss. (1981). Credit rationing in markets with imperfect information. *American Economic Review* 3(71): 393–410.

Stirling, A. (2009). Direction, distribution and diversity! Pluralising progress in innovation, sustainability and development. Available at: *http://mobile.opendocs.ids.ac.uk/opendocs/bitstream/handle/123456789/2458/Direction,%20Distribution%20and%20Diversity.pdf?sequence=1.*

Tullock, G., A. Seldon, and G. L. Brady. (2002). *Government failure: A primer in public choice.* Cato Institute.

Weiss, L. (2008). Crossing the divide: From the military–industrial to the development–procurement complex. In Berkeley Workshop on the "Hidden US Developmental State" (pp. 20–21).

Weiss, L. (2014). *America Inc.: Innovation and enterprise in the national security state.* Cornell University Press.

Wright, B. D. (2012). Grand missions of agricultural innovation. *Research Policy* 41(10): 1716–1728.

Acronyms

ACT	Agreements for Commercializing Technology	LP	Limited partners
AMNPO	Advanced Manufacturing National Program Office	mAbs	Monoclonal antibodies
		MEO	Medium Earth orbit
ARPA-E	Advanced Research Projects Agency-Energy	MII	Manufacturing Innovation Institutes
		NASA	National Aeronautics and Space Administration
CAGR	Compound annual growth rate	NCRA	National Cooperative Research Act
CAN	Cooperative Agreement Notice	NCRPA	National Cooperative Research and Production Act
CAPM	Capital asset pricing model		
CASIS	Center for the Advancement of Science in Space	NOAA	National Oceanic and Atmospheric Administration
CCDev	Commercial Crew Development	NGSO	Non-geosynchronous orbit
COTS	Commercial Orbital Transportation Services	NIH	National Institutes of Health
		NIST	National Institute of Standards and Technology
DARPA	Defense Advanced Research Projects Agency	NL	National Laboratory
		NNMI	National Network for Manufacturing Innovation
DOD	Department of Defense		
DOE	Department of Energy	NSF	National Science Foundation
EAR	Export Administration Regulations	NVCA	National Venture Capital Association
EDF	Electrically defect free	OMB	Office of Management and Budget
FCC	Federal Communications Commission	PPPs	Public-private partnerships
FDA	Food and Drug Administration	PSI	Protein Structure Initiative
FY	Fiscal year	R&D	Research and development
GP	General partner	ROR	Rate of return
GSO	Geosynchronous orbit	SBIR	Small Business Innovation Research
HEO	High Earth orbit	SiC	Silicon carbide
ICT	Information and communication technology	SpaceCom	Space Commerce Conference and Exposition
IP	Intellectual property	STA	Strategic Trade Authorization
IPO	Initial public offering	STEM	Science, Technology, Engineering, and Mathematics
ISS	International Space Station		
IT	Information technology	TOS	Transfer Orbit Stage
ITAR	International Traffic in Arms Regulations	TRLs	Technology Readiness Levels
LBO	Leveraged buyout	VC	Venture Capital
LEO	Low Earth orbit		

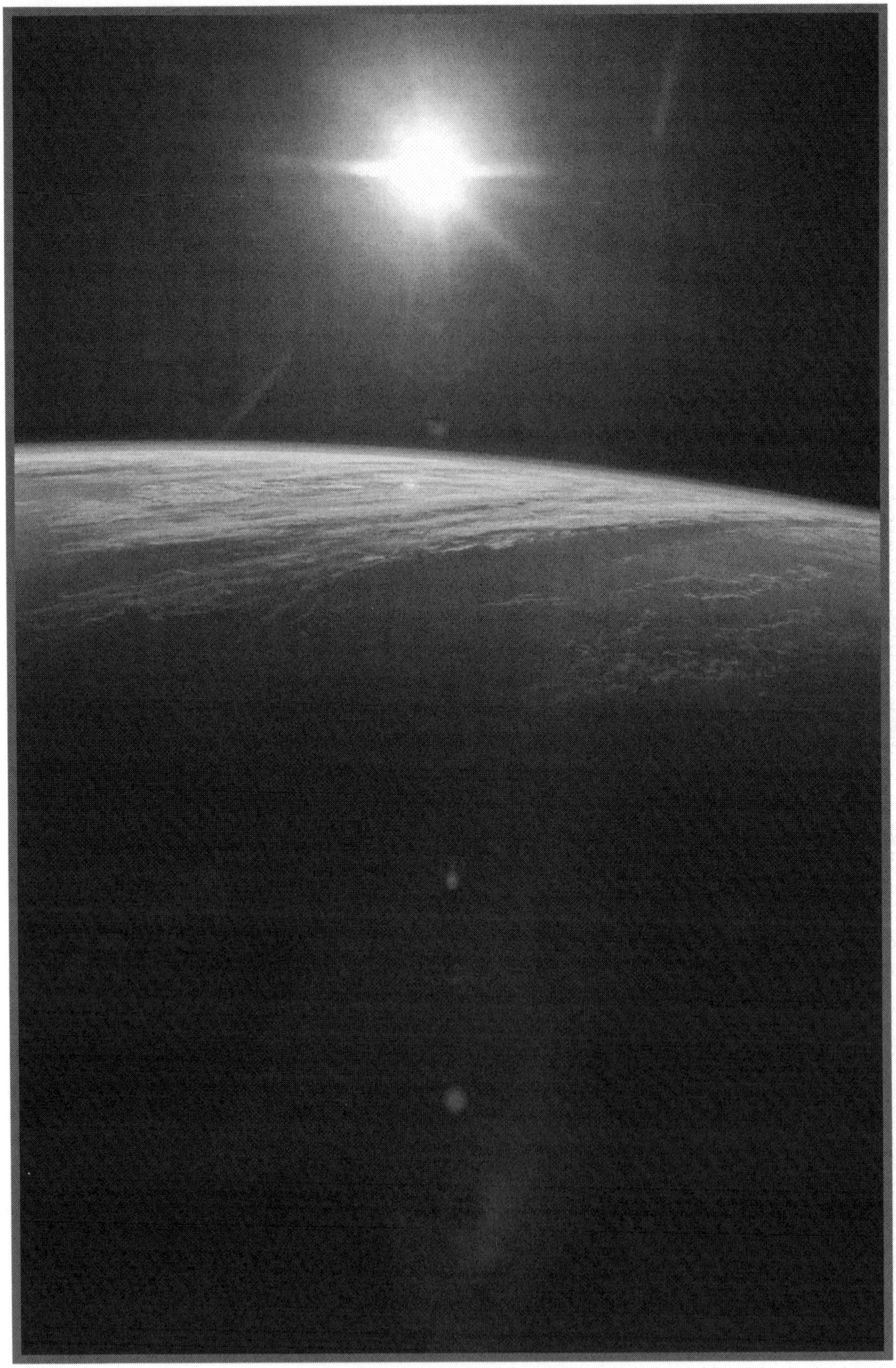

Printed in Great Britain
by Amazon